Lei hao

PERMANENT MAGNET AC MOTOR DRIVE SYSTEM

Lei hao

PERMANENT MAGNET AC MOTOR DRIVE SYSTEM

Analysis, Design and Application

VDM Verlag Dr. Müller

Impressum/Imprint (nur für Deutschland/ only for Germany)
Bibliografische Information der Deutschen Nationalbibliothek: Die Deutsche Nationalbibliothek verzeichnet diese Publikation in der Deutschen Nationalbibliografie; detaillierte bibliografische Daten sind im Internet über http://dnb.d-nb.de abrufbar.
Alle in diesem Buch genannten Marken und Produktnamen unterliegen warenzeichen-, marken- oder patentrechtlichem Schutz bzw. sind Warenzeichen oder eingetragene Warenzeichen der jeweiligen Inhaber. Die Wiedergabe von Marken, Produktnamen, Gebrauchsnamen, Handelsnamen, Warenbezeichnungen u.s.w. in diesem Werk berechtigt auch ohne besondere Kennzeichnung nicht zu der Annahme, dass solche Namen im Sinne der Warenzeichen- und Markenschutzgesetzgebung als frei zu betrachten wären und daher von jedermann benutzt werden dürften.

Coverbild: www.purestockx.com

Verlag: VDM Verlag Dr. Müller Aktiengesellschaft & Co. KG
Dudweiler Landstr. 125 a, 66123 Saarbrücken, Deutschland
Telefon +49 681 9100-698, Telefax +49 681 9100-988, Email: info@vdm-verlag.de
Zugl.: College Station, Texas A&M University, Dissertation 2002

Herstellung in Deutschland:
Schaltungsdienst Lange o.H.G., Zehrensdorfer Str. 11, D-12277 Berlin
Books on Demand GmbH, Gutenbergring 53, D-22848 Norderstedt
Reha GmbH, Dudweiler Landstr. 99, D- 66123 Saarbrücken
ISBN: 978-3-639-07420-8

Imprint (only for USA, GB)
Bibliographic information published by the Deutsche Nationalbibliothek: The Deutsche Nationalbibliothek lists this publication in the Deutsche Nationalbibliografie; detailed bibliographic data are available in the Internet at http://dnb.d-nb.de.
Any brand names and product names mentioned in this book are subject to trademark, brand or patent protection and are trademarks or registered trademarks of their respective holders. The use of brand names, product names, common names, trade names, product descriptions etc. even without
a particular marking in this works is in no way to be construed to mean that such names may be regarded as unrestricted in respect of trademark and brand protection legislation and could thus be used by anyone.

Cover image: www.purestockx.com

Publisher:
VDM Verlag Dr. Müller Aktiengesellschaft & Co. KG
Dudweiler Landstr. 125 a, 66123 Saarbrücken, Germany
Phone +49 681 9100-698, Fax +49 681 9100-988, Email: info@vdm-verlag.de

Copyright © 2008 VDM Verlag Dr. Müller Aktiengesellschaft & Co. KG and licensors
All rights reserved. Saarbrücken 2008

Produced in USA and UK by:
Lightning Source Inc., 1246 Heil Quaker Blvd., La Vergne, TN 37086, USA
Lightning Source UK Ltd., Chapter House, Pitfield, Kiln Farm, Milton Keynes, MK11 3LW, GB
BookSurge, 7290 B. Investment Drive, North Charleston, SC 29418, USA
ISBN: 978-3-639-07420-8

ABSTRACT

Permanent Magnet AC Motor Full Speed Range Operation Using Hybrid Sliding Mode
Observer. (December 2002)

Lei Hao, B.S., Shanghai JiaoTong University, Shanghai, China;

M.S., Institute of Electrical Engineering, Chinese Academy of Science, Beijing, China

Chair of Advisory Committee: Dr. Hamid A. Toliyat

This study focuses on the development and application of full speed range operation of permanent magnet (PM) AC motors (including PM synchronous motor (PMSM) and brushless DC (BLDC) motor) using hybrid sliding mode observers. Specifically, investigations are made under full load starting, sensorless technology and flux weakening control.

The first topic of this study is to analyze the PM spindle motor using finite element analysis (FEA). First the generation of the sinusoidal back-EMF in the spindle motor is investigated. For the sinusoidal PM motor, the condition for producing positive torque is studied when the numbers of stator poles and rotor poles are different. Flux-weakening capability is evaluated in order to prevent the demagnetization of the permanent magnet rotor.

The second topic of this study is to develop an adaptive full-speed range control system of the PM spindle motor considering sensorless technology. Full-load starting and over the rated-speed operations are given more consideration. Adaptive rules are applied to ensure motor performance when motor parameters vary over the full speed range. A hybrid sliding mode observer is developed to estimate the rotor position and speed. MATLAB SIMULAINK is used to verify the basic performances of the overall control system.

Third, a novel full-speed range control system for the brushless DC motor is developed. Unlike the conventional six-step current control, which is said to have poor performance during the flux weakening operation, the developed method can achieve the same controllability as in the sinusoidal PM motor over the flux weakening operation. It can also operate with an increasing motor torque. As the PMSM does, a hybrid sliding mode observer is developed to provide the rotor position and speed. Simulations have been done by using MATLAB SIMULAINK to verify the feasibility of the proposed algorithm.

Finally, experiments have been done at the Whirlpool Corporation to verify theoretical analysis of the proposed PM spindle motor drive system. Texas Instruments digital signal processor, TMS320F2407DSP, is used as the controller. Through the embedded measuring coils and coupled dynamometer, the motor performance over full-speed range has been evaluated. Furthermore, the developed algorithm was put into practice and tested in the prototype Direct-Drive washer.

To my wife, my daughters

and

my parents , my brother

ACKNOWLEDGEMENTS

My deep appreciation goes to many people whose advice, assistance and encouragement have enabled me to complete this dissertation.

I first would like to thank my advisor, Dr. Hamid A. Toliyat. His insight, knowledge and encouragement guided me through my graduate study. His unlimited patience, understanding and willingness to spend his precious time with me are appreciated far more than I have words to express.

I express my sincere gratitude to the members of my graduate study committee, Dr. Prasad Enjeti, Dr. Andrew K. Chan, Dr. Reza Langari, and my substitute Graduate Council Representative, Dr. Charles Taylor, Jr., for serving on my committee and for their invaluable advice and help.

I am also grateful to Dr. Z. Zhang and members of drive systems group of Whirlpool Corporation for their kind guidance and for having made the course of this work especially meaningful.

I also acknowledge the Electrical Engineering department, Texas A&M University, for providing me with the excellent academic circumstances that are essential to the accomplishment of my graduate study. I would like to thank Dr. Chanan Singh, Dr. Garng M. Huang and Ms. Tammy Carda for their help through my years at Texas A&M University.

My thanks are also extended to my fellow colleagues in the Electric Machine and Power Electronics Laboratory: Dr. Shailesh Waikar, Dr. Huangsheng Xu, Dr. Tilak

Gopalarathnam. I cherish the friendship and the good memories I have gained since my arrival at Texas A&M University.

Especially, I am deeply indebted to my wife, Xuelian Zhu. She offered her love and support to help me get to this extraordinary place in my life. I do not have the words to express my gratitude and my love for her.

Last but not least, I would like to extend my appreciation to my parents, my parents-in-law and my brother. No matter how far away from home I am, they are always there to assist me. To my parents, no words can express my gratitude for them for not only providing me endless love and tireless support, but also for teaching me and letting me know about honor, respect and the value of hard work. These have served as the standards by which I measure myself.

TABLE OF CONTENTS

	Page
ABSTRACT	iii
DEDICATION	v
ACKNOWLEDGEMENTS	vi
TABLE OF CONTENTS	viii
LIST OF FIGURES	xii
LIST OF TABLES	xvii

CHAPTER

I	INTRODUCTION	1
	1.1 General	1
	1.2 Literature Review	5
	1.2.1 Motor Analysis	5
	1.2.2 Sensorless Technologies	6
	1.2.3 Flux Weakening Operation	10
	1.3 Research Objectives	12
	1.4 Dissertation Organization	13
II	THE ANALYSIS OF PERMANENT MAGNET SPINDLE MOTORS	15
	2.1 Introduction	15
	2.2 The Calculation of Back-EMF	16
	2.2.1 Magnet Magnetization	19

TABLE OF CONTENTS (Continued)

CHAPTER		Page
	2.2.2 Air Gap Length..	22
	2.2.3 Pole Arc to Pole Pitch Ratio..	23
	2.2.4 The Depth of the Slot..	25
	2.2.5 The Auxiliary Tooth of the Stator...................................	25
	2.2.6 Number of Poles..	28
	2.3 Torque Generation for 36/48 Spindle Motor............................	30
	2.4 The Flux Weakening Capability over Flux Weakening Region....	33
III	HYBRID SLIDING MODE OBSERVER OF PM AC MOTOR...........	40
	3.1 Introduction...	40
	3.2 Theoretical Fundament of Sliding Mode Observer....................	41
	3.3 Hybrid Sliding Motor Observer of PMSM................................	44
	3.3.1 Sliding Mode Current Observer of PMSM.....................	44
	3.3.2 Position Estimation...	46
	3.3.3 Hybrid Sliding Mode Observer of PMSM......................	47
	3.4 Hybrid Sliding Mode Observer of BLDC Motor.......................	50
	3.4.1 Sliding Mode Observer..	51
	3.4.2 Sliding Mode Observer of BLDC motor.........................	52
	3.4.3 Hybrid Sliding Mode Observer of BLDC Motor.............	56
IV	SURFACE MOUNT PERMANENT MAGNET SPINDLE MOTOR FULL-SPEED OPERATION USING HYBRID SLIDING MODE OBSERVER	58
	4.1 Introduction..	58

TABLE OF CONTENTS (Continued)

CHAPTER	Page

	4.2	The Mathematics Model of 36/48 Spindle Motor.............	59
		4.2.1 The Mathematics Model of 36/48 Spindle Motor in the Stationary Frame..	59
		4.2.2 The Mathematics Model of 36/48 Spindle Motor in Rotor Reference Frame..	61
	4.3	Analysis of 36/48 Spindle Motor Drive System................	63
		4.3.1 Analysis of 36/48 Spindle Motor Drive System.........	63
		4.3.2 Optimal Current Control Strategies...................	67
	4.4	Adaptive Full Speed Range Control Algorithm..............	69
		4.4.1 Starting Algorithm.................................	69
		4.4.2 Maximum Torque Control...........................	70
		4.4.3 Flux Weakening Control............................	71
		4.4.4 Adaptive Full Speed Range Control System of PM Spindle Motor..	75
	4.5	Simulation Analysis...	76
V	BLDC MOTOR FULL-SPEED OPERATION USING HYBRID SLIDING MODE OBSERVER		80
	5.1	Introduction...	80
	5.2	Modeling and Analysis of BLDC Machine in Multiple Reference Frame (MRF)..	83
		5.2.1 BLDC Line-Line Model in Stationary Frame............	83
		5.2.2 BLDC Model in Rotor Multiple Reference Frame(MRF)..	87
		5.2.3 Average Model of BLDC Motor in MRF.................	89
	5.3	Full Speed Range Operation Control Algorithm................	92
		5.3.1 Constant Torque Control Algorithm..................	92
		5.3.2 Constant Power Control Algorithm...................	95
		5.3.3 Starting Algorithm.................................	96

TABLE OF CONTENTS (Continued)

CHAPTER			Page
	5.4	BLDC Full Speed Range Control System..........................	97
	5.5	Simulation Analysis……………………………………..	98
VI	SYSTEM IMPLEMENTATION…………………………………..		105
	6.1	Introduction………………………………….......……………	105
	6.2	Experimental Setup…………...…….………………………….	106
	6.3	Hardware Setup of the Control System…………………….….	109
	6.4	Software Development of PM Spindle Motor Control System……	113
	6.5	Experimental Result Analysis…………………………………….	115
		6.5.1 Hall Signal Identification ………..………………………	116
		6.5.2 Starting Process ……………………………………….…	119
		6.5.3 Hybrid Sliding Mode Observer …….…………………….	123
		6.5.4 Maximum Torque/Current Ratio Operation………..……	130
		6.5.5 Flux Weakening Control………………………………....	139
		6.5.6 Washing Machine Application ……………………….…	148
VII	CONCLUSIONS AND FURTHER WORK.. …………………….....		151
	7.1	Conclusions………………………………………………….…	151
	7.2	Suggestions for Further Research……………………………..	154
REFERENCES………………………………………………………………....			155
VITA…………………………………………………………………………..			162

LIST OF FIGURES

		Page
Fig. 1.1	Torque vs speed	3
Fig. 2.1	The geometries of analyzed spindle motors	17
Fig. 2.2	Flux distribution of 36 slots 48 poles spindle motor	18
Fig. 2.3	Flux distribution and back EMF with different magnetization in 3 slot and 4 pole motor	20
Fig. 2.4	Line-to-line back EMF calculated from FEA and measured from experiment	21
Fig. 2.5	Back EMF when air gap is equal to 1mm, 2mm and 3mm	22
Fig. 2.6	The flux distribution (top) and induced back EMF(bottom) vs stator pole arc	24
Fig. 2.7	The flux distribution (top) and induced back EMF(bottom) vs the depth of slot	26
Fig. 2.8	The flux distribution (top) and induced back EMF(bottom) vs the width of slot	27
Fig. 2.9	The normalized flux distribution (a) and induced back EMF(b) vs the number of pole	29
Fig. 2.10	The cogging torque of 36/48 spindle motor from FEA	31
Fig. 2.11	Electromagnetic torque calculated from FEA	33
Fig. 2.12	Flux distribution with and without flux weakening	35
Fig. 2.13	Flux distribution with 45-degree advance angle and without advanced angle	36
Fig. 2.14	Induced voltages in the search coils and their integrate waves during constant torque region	37
Fig. 2.15	Induced voltages of the search coils and their integrate waves during flux weakening region	38
Fig. 2.16	The normalized flux distribution when the motor frequencies are 60Hz, 90Hz, 160Hz, 230Hz, 300Hz and 370Hz	39
Fig. 3.1	The sliding condition	43
Fig. 3.2	Block diagram of the hybrid observer for PMSM	48

LIST OF FIGURES (Continued)

		Page
Fig. 3.3	The location and signals of Hall sensors	49
Fig. 3.4	The relation between e_0 and $e_f = e_3 - \hat{e}_3$	55
Fig. 3.5	Block diagram of hybrid sliding mode observer of BLDC motor	57
Fig. 4.1	3-slot 4-poles part of 36/48 motor	59
Fig. 4.2	The ideal operation of PM spindle motor	65
Fig. 4.3	Optimal control strategies for PMSM	66
Fig. 4.4	Phasor diagram of PM motor considering current and voltage limitation	72
Fig. 4.5	Normalized torque vs speed curves when $x_{qs} I_{as}/E_i$ increases from 0.1 to 1 with 0.1 step	73
Fig. 4.6a	The flow chart of flux weakening control	74
Fig. 4.6b	Block diagram of PMSM drive system	75
Fig. 4.7	Speed response of PMSM	77
Fig. 4.8	Motor actual and estimated q-axis currents in the stationary frame	77
Fig. 4.9	Estimated position from the hybrid observer	78
Fig. 4.10	Q and d axes currents in the rotor reference frame	78
Fig. 4.11	Motor torque and load	79
Fig. 5.1	BLDC motor drive system	81
Fig. 5.2	Constant torque control strategy	81
Fig. 5.3	The ideal back EMF	84
Fig. 5.4	The cross section of BLDC motor	84
Fig. 5.5	The total torque when the back EMF has the different α angles	91
Fig. 5.6	The sum of fifth and seventh harmonics torque when the back EMF has the different α angles	91
Fig. 5.7	Line-to-line motor voltage	94

LIST OF FIGURES (Continued)

Page

Fig. 5.8	The block diagram of full speed range controller	97
Fig. 5.9	Block diagram of BLDC drive system	98
Fig. 5.10	Speed response (commanded and actual speed) in constant torque range	99
Fig 5.11	Load and motor torque	99
Fig.5.12	L-to-L voltage, back-EMF and 10 times the current	100
Fig. 5.13	Commanded speed and actual speed in constant power range	101
Fig. 5.14	Observer output and estimated speed error	101
Fig 5.15	Rotor position and estimated position	102
Fig. 5.16	Estimated position error	102
Fig. 5.17	Fundamental q- and d-axis current in MRF	103
Fig 5.18	Motor torque and load	103
Fig 5.19	Motor line-to-line voltage, back-EMF, and 10*current	104
Fig. 6.1	48-pole 36-slot PM spindle motor	107
Fig. 6.2	Experimental setup of PM spindle drive system	107
Fig. 6.3	DSP controller and inverter	108
Fig. 6.4	Test bench	108
Fig. 6.5	Instruments used for the measurements	109
Fig. 6.6a	Overall hardware architecture of control system	110
Fig.6.6b	The flow chart of the whole control system	114
Fig. 6.7	The Hall sensor signals	117
Fig. 6.8	The line-to-line back-EMF	117
Fig. 6.9	Hall sensor I signal with phase A back-EMF	118
Fig. 6.10	Hall sensor II signal with phase C back-EMF	118

LIST OF FIGURES (Continued)

		Page
Fig. 6.11	Estimation position and Hall signals at 1 rpm	120
Fig. 6.12	Estimation position and Hall signals at 5 rpm................................	120
Fig. 6.13	Starting process using one Hall sensor with 11.3N.m load	121
Fig. 6.14	Starting process using two Hall sensors with 11.3N.m load..............	122
Fig. 6.15	The effects of parameters on sliding mode observer	124
Fig. 6.16	Performance of sliding mode current observer................................	127
Fig. 6.17	Performance of the hybrid sliding mode observer when speed varies from 30Hz to 90Hz	128
Fig. 6.18	Performance of the hybrid sliding mode observer at 30Hz	128
Fig. 6.19	Performance of the hybrid sliding mode observer at 90Hz	129
Fig. 6.20	Performance of the hybrid sliding mode observer at 360Hz	129
Fig. 6.21	Motor dynamic response when the speed changes smoothly with full load	131
Fig. 6.22	Highlight of motor dynamic response when the speed changes smoothly with full load	131
Fig. 6.23	The dynamic response when the speed suddenly varies from 20Hz to 80Hz	132
Fig. 6.24	The dynamic response when the speed suddenly varies from 80Hz to 20Hz	133
Fig. 6.25	The load rejection when 6 N.m load is suddenly added to and removed from the motor	134
Fig. 6.26	The load rejection when 11.3 N.m load is suddenly added to the motor with one-Hall sensor	135
Fig. 6.27	The load rejection when 11.3 N.m load is suddenly added to the motor with two-Hall sensor...	136
Fig. 6.28	Accelerating and decelerating feature under 11.3 N. m load with one-Hall sensor...	137
Fig. 6.29	Accelerating and decelerating feature under 11.3 N. m load with two-Hall sensor...	137
Fig. 6.30	Accelerating and decelerating feature in worse case under 11.3 N.m load with one-Hall sensor..	138

LIST OF FIGURES (Continued)

		Page
Fig. 6.31	Accelerating and decelerating feature in worse case under 11.3 N.m load with two-Hall sensor	138
Fig. 6.32	The system response without flux weakening control while motor speed increases	140
Fig. 6.33	The process of flux weakening control	142
Fig. 6.34	Speed response of flux weakening control with two-Hall sensor under no load	146
Fig. 6.35	Torque, power vs speed	146
Fig. 6.36	The relation of phase voltage and phase induced voltage in flux weakening control	147
Fig. 6.37	Speed response of flux weakening control with one-Hall sensor under no load	147
Fig. 6.38	Direct –drive washing machine	148
Fig. 6.39	The cloth used in the test	149
Fig. 6.40	The high speed operation (900rpm) of D-D washer	149
Fig. 6.41	Speed response of direct drive washer	150
Fig. 6.42	Off-balance operation of direct drive washer	150

LIST OF TABLES

	Page
Table1.1 Application characteristic comparisons..	2
Table 4.1 Boundary of position and excited phase vs Hall state	70
Table 5.1 The normalized amplitude of back- EMF harmonics components.......	90
Table 6.1 Experimental conditions	123

CHAPTER I

INTRODUCTION

1.1 General

Permanent magnet (PM) motor drives have come of age. In general, they can be categorized into two types. The first category called PM synchronous motor (PMSM) drive uses continuous rotor position feedback to feed the motor with sinusoidal voltages or currents, obtained by pulse width modulation of the DC bus. The ideal motor back-EMF is sinusoidal so that a constant torque is produced with the very low ripple when sinusoidal currents flow.

The second category called brushless DC (BLDC) motor drive is based on position feedback that is not continuous, but rather obtained at the fixed points, typically every 60 electrical degree for commutation of the phase currents. Here rectangular-shaped currents are forced into the machine or alternatively, the voltage may be fed in blocks of 120, with just a current limit to ensure that the motor current is held within the motor's capabilities. The ideal back-EMF is trapezoidal. With phase currents timed appropriately with the constant part of the back-EMF, constant torque is generated.

The advents of high performance magnets, like samarium cobalt and neodymium-boron iron, have made PMAC motor drive possible to achieve the performances that can

---

This dissertation follows the style and format of *IEEE Transactions on Industry Applications*.

surpass the conventional DC or induction motor and are becoming more and more attractive for industrial application. As compared with induction motor drives, these possess some distinct advantages such as high power density, high torque to inertia ratio, high efficiency and better controllability [1][2]. Table 1.1 provides a qualitative comparison of the key drive application characteristics for trapezoidal and sinusoidal PMAC motor drives [3]. Induction motor drives are included in this table as a baseline for comparison because of their wide popularity for adjustable speed applications.

Table 1.1 Application characteristic comparisons

	Trapezoidal PMAC	Sinusoidal PMAC	Induction AC
Motor efficiency	+	+	
Torque smoothness		+	+
Open loop control		+	+
Close loop simplicity	+		
Minimum control sensors	+		
Extended speed range		+	+
Motor robustness		+	+

The specific positive characteristics of PMAC machines summarized above usually make them highly attractive candidates for several classes of drive applications such as motion control system, servo-drive, aerospace actuators, Low integral-hp industrial drive, fiber spinning and so on. However, recently two major sectors of consumer market pay more attention to PM motor drive due to these advantages. These two applications are the traction and residential markets, such as electric vehicles and washing machines. A motor drive system in these applications requires high starting torque and long speed range as well as efficiency and power density. This torque requirement is shown in Fig.1.1.

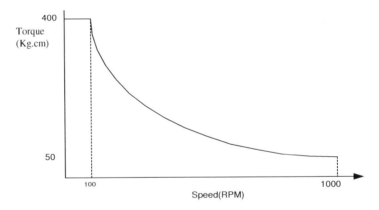

Fig. 1.1 Torque vs speed

In order to satisfy these requirements, the motor should work in two modes. When the motor speed is lower than the rated speed, the motor drive is required to provide constant torque (rated torque). On the other hand, when the speed is higher than the rated speed, the load torque is inverse proportion with the speed. Thus, instead of providing a constant output torque, the constant power operation is preferred because it can significantly reduce the cost and size of the motor drive.

The constant torque operation can be easily achieved in PM motor drives [4][5]. However, when the speed is above the rated speed, the rotational back-EMF of the machine approaches the input source voltage so that the motor drive usually suffers from difficulty to achieve the constant power operation. Recently, flux-weakening control has been developed to deal with this problem. By employing the field component of stator current to weaken the air-gap field produced by permanent magnets, this approach allows some PM motor drives to extend the motor speed.

In PMSM drives, rotor position information is necessary in order to synchronize the phase excitation pulses to the rotor position. This is normally achieved by using position encoders or resolvers, which is mounted against the motor shaft. However these mechanical sensors result in added cost, increased size of the drive, decreased mechanical robustness and reduced reliability of the system. Moreover, it might not allow high-speed operation. Since the customer market is cost-sensitive, sensorless technology should be taken into account for the above two applications.

For same size PM machine, the characteristics of the machine can be varied greatly according to the type of the stator and the rotor. Based on the way the permanent material arranged, basically the rotor can be classified as surface mount magnetic rotor, interior radically- oriented magnetic rotor and interior tangentially-orient magnetic rotor. The stator also has two types according to the winding arrangement: concentrated winding and distribution winding [2][3]. The different combinations of stator and rotor result in the different performances. Therefore, the motor analysis is the first step, also key step, to develop PM motor drive system. Three-slot four-pole PM spindle motor, which usually is used in hard disk drive, is a good choice for traction and residential market because it is easy to manufacture and maintain and has higher power density as well as general advantages of PM motors. Finite element analysis (FEA) is routinely used to evaluate the different types of PM motors.

On the whole, this study mainly focuses on the development and application of the full-speed range PMAC motor drives used in the above two applications. (Notice in this dissertation full speed means starting, low speed to the rated speed and above the rated speed). Specifically, investigations are made under motor analysis using FE methods, full-load starting, sensorless methods and high-speed operation. This approach also can easily be applied to various permanent magnet machines, especially for the widely used, low cost, surface mount, permanent magnet AC machines.

1.2 Literature Review

1.2.1 Motor Analysis

It is well known that sinusoidally excited PM motor has much better controllability than squarewave PM motor in the flux-weakening region. By simply controlling the field component of stator current in rotor reference frame, the air-gap field produced by magnets can be weakened so that the motor speed is extended. At the same time, the motor torque can be controlled separately by adjusting the q-axis component of the current. In addition, torque ripple is associated with the back-EMF in PM motor. Sinusoidally excited PM motor possesses almost ripple-free torque that results in a better performance, for instance, a less noise preferred in the residential market. Therefore, it is desirable to thoroughly analyze the back-EMF of spindle motor and investigate how to generate sinewave back-EMF.

Most of published works on analysis of back-EMF and flux distributions have concentrated on the high-slot stator or treating spindle motor as a special case with large slot opening [6-8]. The analysis of back-EMF in spindle motor has not been totally evaluated.

Since spindle motor usually generates the trapezoidal back-EMF, six-step current control is employed to produce positive torque. The most research on the torque analysis focus on the more precise torque calculation in order to reduce torque ripple [9-13]. None of them investigate how to inject the sinusoidal current in spindle motor such that a positive torque can be generated, especially if the stator rotor number of poles are different.

Another important problem for motor analysis is the flux weakening capability, which is necessary for the extension of speed without demagnetizing magnet. Since spindle motor seldom operates in flux weakening region, it is an untouched area.

1.2.2 Sensorless Technologies

Various methods have been presented in the literature for detecting the rotor position of different classes of permanent magnet motors. Majorities of the methods have been developed for the salient rotor permanent magnet motor. The variations in the d and q axis inductances give a parameter that could be monitored to detect the rotor position. [14-16]. However, a few papers have been published on the surface-mount non-salient permanent magnet synchronous motors (PMSM) since the non-salient rotor has been the equal inductances along both d and q axes. In this section, since the spindle motor is the surface mount PM motor, only the techniques for non-salient PMSM are briefly explained according to the following five categories [17]:

a) Those using back-EMF method.
b) Those using measured current, voltage, fundamental machine equation, and algebraic manipulations.
c) Those using observers.
d) Sensorless starting techniques.
e) Those with novel techniques not falling into previous four categories.

Wu and Slemon [18] used the back-EMF estimated from the measurement of terminal voltages and stator currents. The rotor flux quantities are estimated by the integration of the machine terminal voltages and the rotor angle information was obtained from these quantities. This method suffered from limitations at low speed because the back-EMF generated is proportional to the speed, and at low speeds the back EMF is very low and is buried in the switching noise of the PWM inverter driving the motor.

For those position signals based on the measurement of the voltage and currents, [19-21] initially the measured voltages and currents are transformed to rotor and stator d-q frames. Relations between stator and rotor reference frames allow the substitution of the stator reference frame quantities into voltage equation in the rotor reference frame. Once the substitution are made and the equation relating rotor and stator references are totally in terms of the stator variables, which are then manipulated to yield rotor angle. However, these methods usually suffer from the poor precision and offset problem of the integral algorithm and the low accuracy of differentiates operation as well as the low speed operation due to the same reason as the back –EMF methods.

Observer methods have received an increased interest in recent years and are preferred for medium to high-speed operations. Extended Kalman filter (EKF) is used as a position estimator in [22-23]. Though EKF is well known, the convergence of the estimated positions is difficult to be guaranteed and computational complexity is a detrimental factor in its widespread use. Other observer systems include non-linear, full-order, disturbance, reduced order, sliding mode and etc [24-29]. The sliding mode observer with the potential advantages similar to those of the sliding controller such as robust can be expected to be an important candidate for the state estimator in motor position and speed. However, the lack of initial information of the state is a drawback for these observer-based techniques. It results in deteriorating performance at zero and around low-speed operation. Accumulating error is another problem.

Several studies focus on the standstill and low speed operation using signal injection based on magnetic saliency [30-32]. However, these techniques work well when interior permanent magnet (IPM) motors since inherent rotor saliency are used. For surface mount

PMSM, it is difficult to find the d and q axis because both d and q axis are saturated when the motor is loaded as well as the complexity.

There are several sensorless methods, which do not belong to above fields [33-34]. Instead of using standard PM motors, Nondahl, Ray and Schmidt presented a method which obtains the information of the rotor position by modifying motor [33]. By adding a conducting winding that links only the d-axis of the rotor, the stator winding inductance becomes different when the winding is aligned with d-axis and q-axis. The variation in stator winding inductance can be used to determine the rotor position. However, the error of the position estimation increases with motor load. Toliyat proposed a method to detect rotor position by using the taps on the stator [34]. These taps were used to measure the voltages induced in the coils of the stator. The difference between the two tap voltages was taken and the resultant was used to estimate the position. It was shown that the resistive component of the voltage was cancelled and most of the inductive drops were also cancelled. This method is suitable for very low speed but fails at standstill.

Low-cost Hall-Effect sensors (<$2) are introduced to assist a simplified speed observer based on the mechanical motion equation in [35-36]. It has a good performance in very low speed but the error in high speed is not acceptable.

In all, none of the sensorless technology for surface mount PMSM can be suitable from standstill to high speed under full load condition.

1.2.3 Flux Weakening Operation

Since the conventional d-q model can be directly applied to the sinewave PM motor drive, a number of papers have been published to investigate the flux-weakening operation of PMSM. The early works only deal with the analysis and control of the existing machines [37-38].

Schiferl [39] introduced the per-unit system to reduce the motor parameter and showed the motor drive design criterion for optimum flux-weakening performance. Morimoto [40] analyzed the infinite maximum speed of PM motor using circle diagram and described the optimum current control strategy. He also investigated demagnetization limitation. Adnanes[41-42] normalized the d-q model to per-unit base model and investigated the flux-weakening performance of the finite maximum speed drive. Soong [43-44] summarized all above works and graphically illustrated the effect of varying the drive parameters on the shape of the optimal flux-weakening characteristic of PM motor by using parameter plane. He also examined the effect of the resistance, saturation and iron loss on the performance of the optimal flux-weakening operation. All above works set forth the theoretical principle for flux-weakening operation of PMSM.

R.Krishnan [45] addressed a method which keeps the motor at its maximum speed-torque envelop within the inverter's maximum current and voltage limits. However, overload condition has not been taken into consideration. Bianchi [46] investigated the effects of currents, voltages and torque of different drive failures occurring under flux-weakening operation. Those approaches provide a starting point to design a practical flux-weakening motor drive system.

The above reviews outline the flux-weakening operation of PMSM. On the other hand, very few efforts have been done for BLDC motor because d and q axis currents can't be decoupled with trapezoidal BEMF. Thus, phase advance-angle control remains the flux-weakening technique in BLDC motor drive [47]. This method has been proved useful in many cases. However, the drawback is quantitative speed or torque control as compared with PMSM drive in flux-weakening region can not be obtained. In other words, considering some speed or torque point, how much advance-angle should be used is unknown. Furthermore, since the quasi-square wave (six-step) current control, which at any time only two phases are conducting and each phase conducts 120 degree in positive and 120 degree in negative directions, is used to generate the positive torque, the capabilities of inverter and motor have not been totally exploited.

The alternative methods [48-49] extend the motor speed range by using the stator excitation applied over 180 degree, rather than 120 degree. However, it still has not obtained quantitative speed or torque control.

The poor performances of BLDC motor in flux weakening region limit the applications of BLDC motor in the high-speed range. Since the intersection between squarewave current and squarewave magnetic field in the motor can produce a larger torque than that produced by sinewave current and sinewave magnetic field, it is desirable for traction and washing machine applications to develop a quantitative flux weakening control algorithm for BLDC motor drive.

1.3 Research Objectives

Based on the above discussion. This research focuses on the following areas:

The first objective of this study is to analyze the PM spindle motor using FEA. How to generate the sinusoidal back-EMF in spindle motor is investigated initally. Second, for sinusoidal PM motor, the development of the positive torque is exploited if e numbers of stator poles and rotor poles are different. Finally flux-weakening capability is evaluated in order to prevent demagnetization of magnet.

The second objective of this study is to develop an adaptive full-speed range control system of PM spindle motor considering the sensorless technology based on the knowledge of PMSM. Full load starting, very low speed and above the rated speed operation are paid more attention. Adaptive rules are applied to ensure the motor performance when the motor parameters vary over the long speed range. A trade-off hybrid sliding mode observer is developed to estimate the rotor position and speed.

Thirdly, a novel full-speed range control system for Brushless DC motor is developed. Unlike the conventional six-step current control, which is argued to have poor performance during flux weakening operation, the proposed method can achieve the same controllability as in the sinusoidal PM motor over the flux weakening operation as well as the increasing developed torque. As the PMSM does, a hybrid sliding mode observer should be developed to provide the rotor position and speed.

Finally, experiments for the developed algorithms of PM spindle had been performed at the Whirlpool Corporation by using TMS320F2407 DSP to verify the theoretical analysis and simulation results. Through the embedded measuring coils and coupled dynamometer, the motor performance over full-speed range is evaluated. Furthermore, the developed algorithm was put into practice and tested in the prototype Direct-Drive washer.

1.4 Dissertation Organization

This dissertation consists of seven chapters. After a brief introduction and literature review of PMAC motor drive in Chapter I, Chapter II investigates several important characteristics of PM spindle motor. These include sinusoidal back-EMF generation, toque production and flux weakening capability.

Chapter III develops the hybrid sliding mode observers for both PMSM and BLDC motor, which provide the position and speed signals to close the speed loop and to synchronize the phase excitation pulses to the rotor position.

In Chapter IV, an adaptive full-speed range control system of the PM spindle motor using hybrid sliding mode observer is developed. The full-load starting, maximum torque/current ratio current control and practice flux weakening control are detailed. The simulation results are also included in this chapter.

Chapter V discusses a novel full-speed range control system of the BLDC motor. Instead of conducting the two switches at every moment, three switches conduct, which results in an increasing developed torque. Also, the developed algorithm shows the same controllability as in the sinusoidal PM motor over the flux weakening operation compared to the conventional six-step BLDC drive system. The algorithms are explained and the simulation results are provided to verify the validity of the proposed method.

In Chapter VI, the experimental setup and results are presented. A series of tests are implemented for the PM spindle motor.

Conclusions are made in the last Chapter – Chapter VII. Also suggested further research are proposed.

CHAPTER II

THE ANALYSIS OF PERMANENT MAGNET SPINDLE MOTORS

2.1 Introduction

In PM motors, the air gap magnetic field is produced by the rotor permanent magnet material. Though it provides an unique advantage for designer to design different geometries for different applications, on the other hand, the motor analysis becomes necessary for the successful design of the PM drive system since it is associated with the structure of motor. Therefore, the first step of the design of the PM drive system is the analysis of PM motor, which is the objective of this chapter.

As mentioned in Chapter I, the sinusoidally excited PM motor has much better controllability in the flux-weakening region and better performance for the entire speed range, which is preferred in the residential market, than the squarewave PM motor. Hence, the first topic of this chapter is to investigate how to generate sinewave back-EMF in spindle motors. Since the back-EMF can be calculated by the derivative of the flux linkage over effective tooth area during 360 electrical degree period for spindle motor, both flux distribution and back-EMF are discussed in this chapter.

Conventionally, the six-step current control is used for spindle motor control. After producing the sinusoidal back-EMF, the next problem for spindle motor becomes how to generate positive torque by injecting sinusoidal currents. It is the key issue for the whole drive system and also the new topic for PM spindle motor. The solution of this problem is the second topic in this chapter.

Finally, the whole drive system is required to run over a very long speed range under flux weakening control. Thus, the analysis of the flux density in flux weakening region, which is important for flux weakening operation in order to prevent the demagnetization of magnet, should been taken into account for the analysis of spindle motor. It is also a new topic for spindle motor.

According to above discussions, the objectives of this chapter focus on the three topics. Section 2.1 investigates the back-EMF. Section 2.2 discusses the torque generation and section 2.3 deals with the flux weakening operation.

2.2 The Calculation of Back-EMF

In this section, the prototype spindle motor, 36/48 motor (slot/pole=3/4) with outer rotor, is analyzed and some results are compared with those of 3/4 (slot/pole) series motor in order to investigate the effects of the parameters of motor structure in generating sinusoidal back-EMF. It is assumed that both motors have the same maximum flux density in tooth, stator core, and rotor core. Fig. 2.1 shows the geometries of these motors. For 3-slot 4-pole motor, the rotor pole is divided into 12 parts. Each part has the same size as the rotor pole of 36-slot 48-pole motor.

In a spindle motor, the back-EMF can be calculated from the single tooth flux [8]:

$$e_T = -\omega \frac{d\psi_T}{d\theta} \tag{2.1}$$

where ψ_T is the flux linkage over the effective tooth area (1/3 high form stator core), ω is the speed and θ is the rotor position in electrical degree. It is assumed that the motor speed is constant at its rated value. The effective tooth area is the same for same number of poles of the motor. Hence, the calculation of back-EMF in turn can be the calculation of flux distribution. Fig.2. 2 shows the flux distribution of the 36/48 spindle motor.

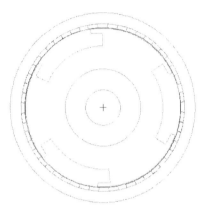

(a) 3 slot and 4 pole outer rotor motor

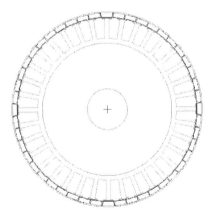

(b) 36 slots and 48 poles outer rotor motor

Fig. 2.1　　The geometries of analyzed spindle motors

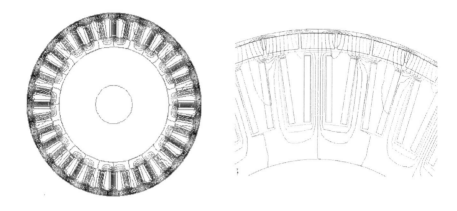

Fig. 2.2 Flux distribution of 36 slots 48 poles spindle motor

According to [7], the flux distribution in smooth stator can be affected by the following parameters:

1) magnet magnetization,

2) air gap length,

3) pole arc to pole pitch ratio,

4) number of poles

In spindle motors, some of these parameters are not affected much and some new parameters should be added. These are:

1) slot depth,

2) slot width (similar to adding the auxiliary tooth in stator),

3) pole number (pole/slot ratio keeps constant).

In this section, the effects of the first four parameters are investigated and the results are compared with smooth stator. In addition, the last three special cases for spindle motor are evaluated.

2.2.1 Magnet Magnetization

Usually there are two types of magnet magnetization: parallel and radial magnet magnetization. For smooth stator, parallel magnet magnetization generates more near sinusoidal air-gap flux distribution and radial magnet magnetization produces near rectangular distribution [50]. For spindle motor, the differences due to the magnet magnetization are not significance. Fig.2.3 demonstrates the flux distribution and induced back-EMF of 3 slot, 4 pole motor with parallel and radial magnet magnetizations. Both of them almost hold similar shapes. The reason is that the differences of the shapes due to the effects of magnetization near the arc fringe are much reduced by the large slot opening in spindle motor.

As the number of pole increases, each pole occupies less and less degree of overall 360 mechanical degree of the rotor. If the rotor diameter is large, both of methods of magnet magnetization behave close to each other, which results in the similar waveform. Hence, herein it is assumed that the two kinds of waveforms due to different magnet magnetization are the same in 36/48 motor. Fig.2.4 shows back-EMF of the 36/48 spindle motor calculated from FE simulation and the experimental result respectively. It can be seen that simulation and experimental results match each other and have a sinusoidal shape. The reasons to generate the sinusoidal back-EMF are the main objective of this section.

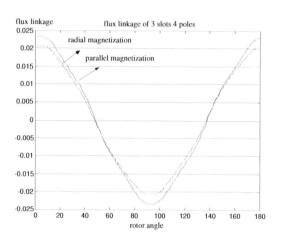

(a) flux distribution

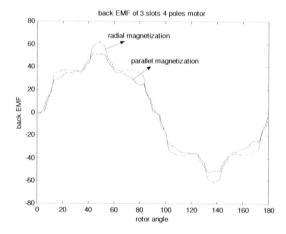

(b) back EMF

Fig. 2.3 Flux distribution and back EMF with different magnetization in 3 slot and 4 pole motor

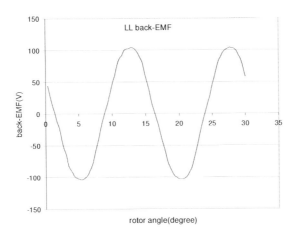

(a) Line-to-line back-EMF calculated from FEA

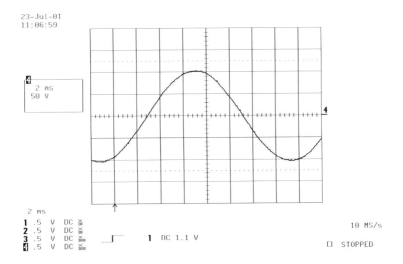

(b) Line-to-line back –EMF measured from the experiment

Fig. 2.4 Line-to-line back EMF calculated from FEA and measured from the experiment

2.2.2 Air Gap Length

Fig. 2.5 shows the back-EMF of 36/48 motor when the air-gap length varies. It can be seen that the waveforms have similar shapes. The reason is that the leakage flux around the edge of magnet, which results in the variation of the shape of the flux distribution in smooth stator motor, changes very little due to large slot opening in spindle motor even when the air gap increases. The increase of air gap only results in the smaller amplitude of flux density in the air gap.

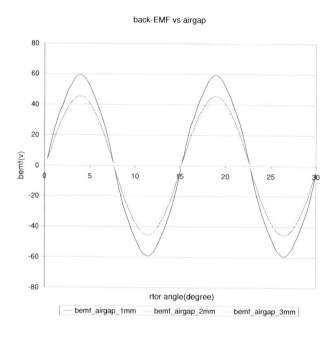

Fig. 2.5 Back EMF when air gap is equal to 1mm, 2mm and 3mm

2.2.3 Pole Arc to Pole Pitch Ratio

Conventionally, spindle motor has the trapezoidal flux distribution. In trapezoidal PM motor, the pole arc to pole pitch ratio should be chosen as large as possible in order to minimize the torque ripple. However, since the objective of this section is to analyze the effect of parameters in generating the sinusoidal back-EMF, the effect of varying the pole arc to pole pitch ratio has been evaluated. It is known that the short-pitched distribution winding can reduce the harmonics, which results in a more sinusoidal flux distribution for smooth rotor. Therefore, pole arc to pole pitch plays an important role in generating sinusoidal waveform for distribution winding.

For proposed spindle motor, the stator structure is the same. Thus, when pole arc to pole pitch ratio changes, it is similar to the stator winding pitch variations from short-pitched to long-pitched. Fig. 2.6 shows the flux distribution and induced back-EMF when stator's winding pitch changes from short-pitched to long-pitched. It is clear that the shape of waveform becomes more flat and the amplitude decreases when the stator winding pitch increases. However, though short pitched reduces some harmonics and the waveform become more sinusoidally distributed, the full pitched waveform is still good sinusoidal in shape. Hence, the short pitched does not mainly contribute to the generation of sinusoidal back-EMF.

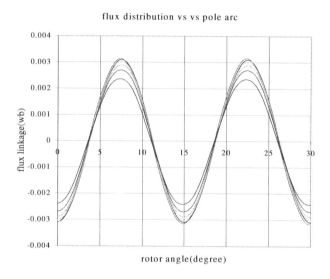

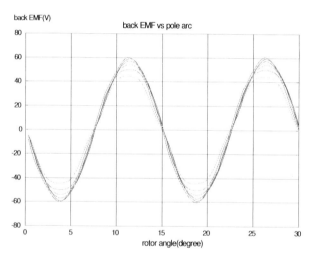

The stator-winding pitch /pole arc are 0.818, .0.9,1.0,1.125,1.286 and 1.5
when the amplitudes change from high to low

Fig. 2.6 The flux distribution (top) and induced back EMF(bottom) vs stator pole arc

2.2.4 The Depth of the Slot

In case of spindle motor the depth of slot affects the shape of the air gap flux distribution. The FE results are demonstrated in Fig. 2.7. The depth of slot changes the leakage flux when it becomes smaller. Thus, the curves become more flat than sinusoidal when the stator is more close to smooth stator and the waveform is the same if the depth of slot is more than 1/2 the depth of the prototype motor.

2.2.5 The Auxiliary Tooth of the Stator

The auxiliary tooth introduces the new flux path in spindle motor that results in the variation of air gap flux distribution. The results are shown in Fig. 2.8. It can be seen that the back EMF becomes more flat than the prototype motor when the auxiliary tooth is employed. If the width of auxiliary tooth increases similar to narrow slot width, until smooth stator is obtained, the amplitude of back-EMF reduces. It is easily understood because some main fluxes become the leakage fluxes.

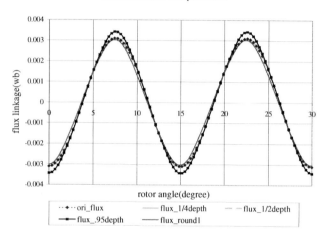

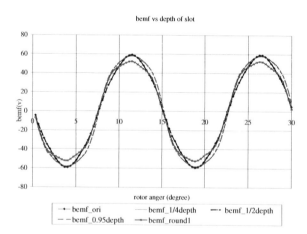

The depth of slot are 0(smooth stator), 1/4, 2/1, 0.95 and 1(original stator) times depth of slot of prototype motor

Fig. 2.7 The flux distribution (top) and induced back EMF(bottom) vs the depth of slot

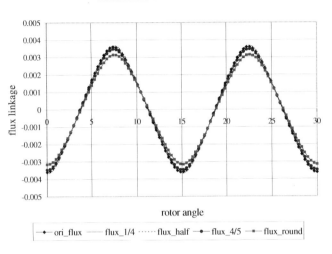

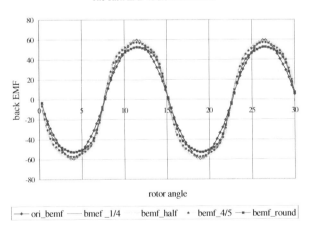

The width of auxiliary tooth is 0(original stator), 1/4, 2/1, 4/5 and 1(smoth stator) times width of tooth of prototype motor

Fig. 2.8 The flux distribution (top) and the induced back EMF(bottom) vs the width of slot

2.2.6 Number of Poles

In 3/4 series motor, there are many pole slot combinations such as 3/4, 6/8, 12/16, 18/24, 36/48 and so on. The different combination gives the different shape of back-EMF even though the tooth, stator core, and rotor core have the same saturation levels for different combinations. Fig.2.9 illustrates these results. The 3-slot, 4-pole motor has the six-step back-EMF which is well known in 3/4 series motor. As the number of pole increases, the shape of back-EMF changes and becomes more like sinusoidal waveform. After the number of poles is above certain value (48 in this case), the curve becomes more sinusoidal. The reason may be the calculation of flux density in the air-gap is related to the flux path. When the number of poles increase, the flux path in rotor and stator core varies from arc to straight line and becomes smaller so that the back EMF curve changes from six-step to smooth curve.

Combining the effect of the pole number, the depth of slot and the width of slot, the increase in the number of pole causes the back-EMF to be a smooth curve and close to a sinusoidal. The large slot opening shape it to a more sinusoidal curve.

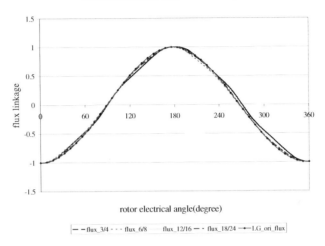

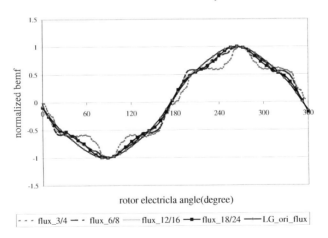

The combination of slot and pole is 3/4, 6/8, 12/16, 18/24 and 36/48

Fig. 2.9 The normalized flux distribution (top) and induced back EMF(bottom) vs the number of pole

2.3 Torque Generation for 36/48 Spindle Motor

For a PMAC motor, torque consists of electromagnetic and reluctance components. The reluctance component is produced by the magnetic attraction between the rotor-mounted permanent magnets and the stator teeth and can be given by [9]:

$$T_{cog} = \sum_{1}^{P}(-\frac{1}{2}\phi_g^2 \frac{d\mathfrak{R}}{d\theta}) \tag{2.2}$$

where: P is the number of pole, $\mathfrak{R}$ is the reluctance and ϕ_g is the air gap magnetic flux.

It is the circumferential component of attractive force that attempts to maintain the alignment between the stator teeth and the permanent magnets. The average value for one circle is equal to zero. Thus, it just affects the motor performance and does not contribute to average torque. Since the applications discussed herein always start with the high load, which filters the torque ripple, the effect of cogging torque can be ignored. Fig 2.10 shows the cogging torque of 36/48 spindle motor from FEA.

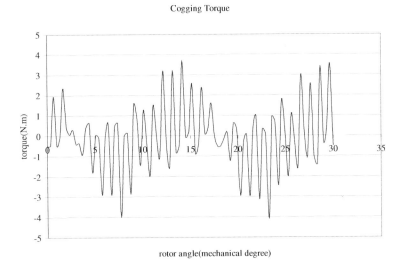

Fig. 2.10 The cogging torque of 36/48 spindle motor from FEA

Another torque component is the electromagnetic torque. It is produced by interaction of stator and rotor magnetic field and provides the positive average torque in PM motor. Typically six-step current control is used to produce the electromagnetic torque by injecting squarewave current. Since the proposed spindle motor has the sinusoidal back-EMF, sinusoidal current should be injected to obtain positive average electromagnetic torque. If the knowledge of conventional PMSM is employed, the rotor pole and stator pole should be the same to produce the positive torque. However, the proposed motor has 36 slots in the stator and 48 poles in the rotor. It is a three-phase motor. Since one slot is one phase, the stator winding only can provide 24 poles. Therefore, how to generate the positive torque with 24 poles in the stator and 48 poles in the rotor becomes a challenge.

To solve this problem, first the conventional four-pole motor is selected as a reference. The difference of these two motors are that spindle motor has three winding set and one winding set is one phase, and conventional motor has six winding set, which two winding sets can be connect in parallel or series and then become a three-phase motor. If three-phase currents are applied to the stator winding, two parameters should be considered: one is the frequency and another is phase shift between two phases. Assuming 60 Hz three phase currents are applied to the stator windings of conventional motor, rotor will run at 30 Hz mechanical frequency and 60 Hz electrical frequency. For spindle motor, the situation should be the same. If the rotor run in 30 Hz mechanical frequency, stator currents should have a frequency equal to 60 Hz. However, if the electrical degree is considered for phase shift, it can be found that there is 120-phase shift between the two phases in conventional motor and 240-phase shift between the two phases in the proposed motor. Thus, Combining the two parameters, it is possible to generate positive torque for spindle motor. Instead injecting 120-degree phase shift currents, 240-degree phase shift currents are used in the spindle motor. These three-phase currents are also three balance currents.

The three-phase currents that used in the FEA are defined as:

$$\begin{cases} i_{as} = I_s \sin(24 \cdot \omega_m \cdot t) \\ i_{bs} = I_s \sin(24 \cdot \omega_m \cdot t - 240) \\ i_{cs} = I_s \sin(24 \cdot \omega_m \cdot t + 120) \end{cases} \quad (2.3)$$

where ω_m is the mechanical angular velocity.

The Fig 2.11 shows the electromagnetic torque calculated from FEA. The average value is 13.77 N. m. The rated torque obtained from data sheet of the manufacture is 13.4N.m. These two numbers are very close.

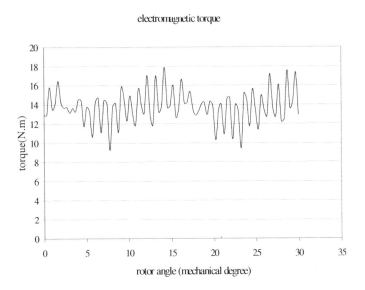

Fig. 2.11 Electromagnetic torque calculated from FEA

2.4 The Flux Weakening Capability over Flux Weakening Region

In electrical vehicles and washing machines, when the speed is higher than the rated speed, the load torque decreases when the speed increases. Instead of providing the constant torque operation, constant power operation is preferred in order to reduce the cost and size of motor drive. Therefore, it is necessary to investigate the flux distribution in the flux weakening region of PM motor in order to explore the flux weakening capability of the prototype motor and avoid demagnetization.

The FE analysis has been done to predict the variation of flux distribution during flux weakening region. Advanced angle control is employed to achieve the flux weakening operation. Fig. 2.12 shows the flux distribution without advanced angle and with 45 degree advanced angle. It can be seen that the flux linkage becomes less during flux weakening region. Fig. 2.13 demonstrates the value of flux distributions compared to experimental results, which match each other.

Experiments have been done to investigate the flux distribution in whole operation region. In order to perform the task, search coils are added in the stator tooth and stator core to measure the induced voltages. The flux distributions can be obtained by integrating the induced voltages in these search coils. Field-oriented control is employed to run the motor over the constant torque region and the constant power region. The flux weakening control starts from 90 Hz until 380 Hz. The amplitude of phase current and DC bus voltage stay constant over the whole speed range. Fig 2.14 displays the measured voltage in the search coil and integrated flux distribution in the air-gap and stator core when motor runs in constant torque region. It can be seen that the flux distribution in the air gap stays sinusoidal and the distribution in stator core becomes flat due to the saturation. Fig. 2.15 demonstrates the results of the flux weakening operation. The amplitude of flux linkage reduces and the flux distribution in stator core departs from saturation with the increase in speed. Fig. 2.16 shows the normalized flux distribution in the flux- weakening region. The magnetic material is still in the range of recovery until motor frequency equals 370 Hz since the amplitude of the flux linkage almost decreases linearly with the increase of the speed. Hence, it is safe to extend motor speed up to 900 rpm (corresponds to 360 Hz) without demagnetizing the permanent magnet material.

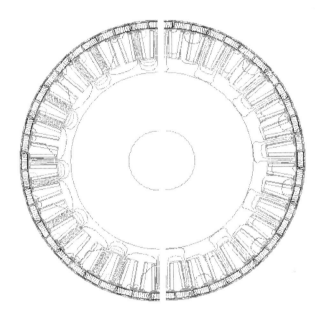

without advanced angle(left), with 45 advanced angle (right)

Fig. 2.12 Flux distribution with and without flux weakening

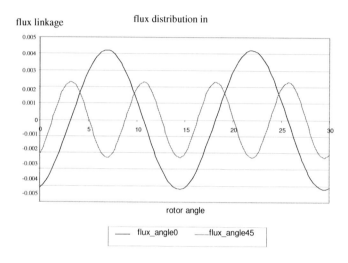

(a) flux distribution obtained from FEA

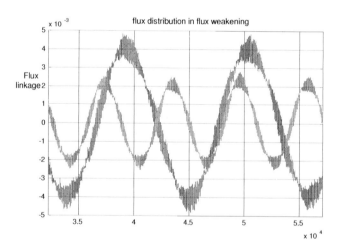

(a) flux distribution obtained from experiment

Fig 2.13 Flux distribution with 45-degree advance angle and without advanced angle

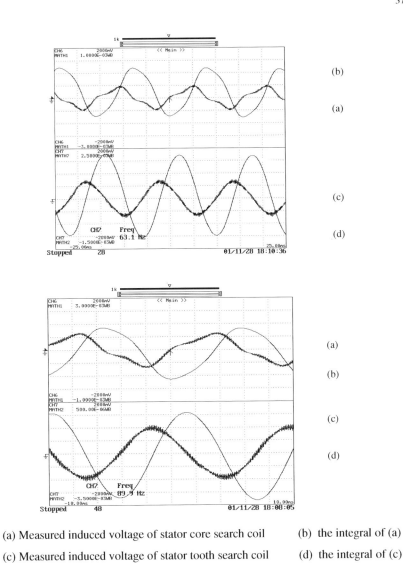

(a) Measured induced voltage of stator core search coil   (b) the integral of (a)
(c) Measured induced voltage of stator tooth search coil   (d) the integral of (c)

Fig. 2.14 Induced voltages in the search coils and their integrate waves during constant torque region

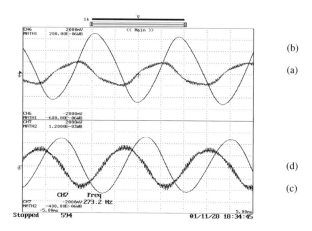

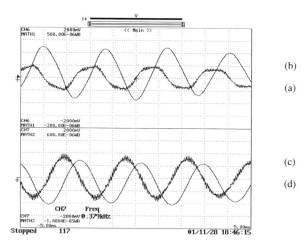

(a) Measured induced voltage of stator core search coil (b) the integral of (a)
(c) Measured induced voltage of stator tooth search coil (d) the integral of (c)

Fig. 2.15 Induced voltages of the search coils and their integrate waves during flux weakening region

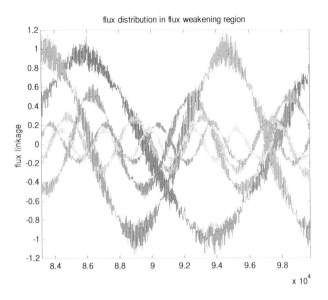

Fig 2.16 The normalized flux distribution when the motor frequencies are 60Hz, 90Hz, 160Hz, 230Hz, 300Hz and 370Hz

CHAPTER III

HYBRID SLIDING MODE OBSERVER OF PM AC MOTOR

3.1 Introduction

Since the PM motors discussed herein are specific examples of synchronous motor, average torque can be produced only when the excitation is precisely synchronized with the rotor frequency and instantaneous position. The most direct and powerful means of ensuring that this requirement is always met is to continuously measure the rotor's absolute angular position such that the excitation can be switched among the PM motor phase in exact synchronism with the rotor's motion.

There are two types of methods to provide the required rotor position: one group use the position sensors, such as encodes and resolvers for sinusoidal PM motor and Hall sensor for trapezoidal PM motor; another group employ the sensorless techniques. Though the mechanical sensors can satisfy the requirement of position information, these result in added cost, increased size of the drive, decreased mechanical robustness and reduced reliability of the system. On the other hand, though sensorless techniques eliminate the disadvantage of position sensor, these methods suffer from starting and very low speed under full load condition.

In this chapter, the hybrid sliding mode observers with Hall effect sensors are developed for both PMSM and BLDC motor. These two observers have the same structure, which consists of two parts: a simple observer and a sliding mode observer with Hall-sensor correction. A simple speed and position observer based on Hall sensor signals is employed for standstill and low speed operation. The sliding mode observer starts to work from the stand still and replaces the simple observer from medium speed at which the sliding mode observer is already converged. The estimated position is corrected by Hall sensor signals. Thus, this hybrid observer has all the advantages of sliding mode observer and improves the zero and low-speed operations, which is the main drawback in observer based methods. Furthermore, the accumulating error and phase shift are avoided because the estimated position is corrected every 180/n (n is the number of Hall sensors) electrical degrees.

This chapter is organized as follows: Section 3.2 presents the fundament of sliding mode observer theory; A hybrid sliding observer for PMSM is discussed in section 3.3; Section 3.4 explains the hybrid sliding mode observer designed for BLDC motors.

3.2 Theoretical Fundament of Sliding Mode Observer

In sliding mode observer, the sliding problem is treated as a special case of a state controller problem [51]. Thus, the sliding mode control is first discussed before explaining sliding mode observer.

Let a nonlinear system defined by differential equations in n-dimensional state space with m-dimensional control action be:

$$\dot{x} = f(x,t,u) \tag{3.1}$$

where $x \in \Re^n$ are the states and $u \in \Re^m$ is the control signal. Then, the design procedure of sliding mode consists of two steps. First, a sliding surface, $s(x;t) \in \Re^n$ is selected such that the desired system dynamics are achieved according to some performance criterion such as tracking, regulation and stability, when the system state trajectories are restricted to the sliding surface. Second, a discontinuous control law u(x; t) is defined such that the states of the system reach the sliding surface and sliding mode exists on this surface[52].

A time varying sliding surface S(t) is first defined by the scalar equation $s(\tilde{x};t) = 0$ with

$$s(\tilde{x};t) = (\frac{d}{dt} + \lambda)^{n-1} \tilde{x}, \quad \lambda > 0 \tag{3.2}$$

where λ is a positive constant and $\tilde{x} = x - x_d$ is the tracking error vector. The problem of tracking $x \equiv x_d$ is equivalent to force the state x of a plant remain on the surface S(t). When $s(\tilde{x};t)$ is greater than zero, the structure is varied such that the states, manipulated by control law, decrease $s(\tilde{x};t)$, and the opposite occurs when $s(\tilde{x};t)$ is less than zero. The sufficient condition for such positive invariance of S(t) is to choose the control law such that outside of S(t)

$$\frac{1}{2}\frac{d}{dt}s^2(\tilde{x};t) \leq -\eta|S| \tag{3.3}$$

where η is a positive constant. The above inequality constraints trajectories to point towards the surface S(t) as shown in Fig.3.1. The surface S(t) is called sliding surface, and the condition of control is known as the sliding condition.

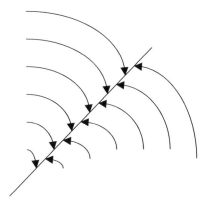

Fig.3.1 The sliding condition

The problem in similarly exploiting sliding behavior in the design of observer, rather than controllers, is that the full state is not available for measurement, and the sliding surface definition is not adequate. Thus, in the design of the sliding mode observer, the sliding problem is treated as a special case of a state controller problem, where the sliding surface are designed based on the error dynamics of the state variable estimates. The sliding surface is reached when the state estimation error goes to zero in a finite time and at the same time all other errors in the state estimates decay exponentially. Based on the above idea, the sliding mode observer of both PMSM and BLDC motor are developed in the following section.

3.3 Hybrid Sliding Motor Observer of PMSM

3.3.1 Sliding Mode Current Observer of PMSM

Since only the terminal quantities such as voltage or current on the stationary reference frame are the available measured variables for PMSM, the sliding mode current observer is built based on PMSM model in the stationary reference frame. The PMSM model in the stationary reference frame is characterized by:

$$\frac{d}{dt}i_s = A \cdot i_s + B \cdot v_s - B \cdot E_s \tag{3.4}$$

where $i_s = [i_{ds}^s \quad i_{qs}^s]$, $v_s = [v_{ds}^s \quad v_{qs}^s]^T$, $E_s = [e_{ds}^s \quad e_{qs}^s]$

$$A = (-r_s / L_s) \cdot I \tag{3.5}$$

$$B = (1 / L_s) \cdot I \tag{3.6}$$

$$\begin{cases} e_{ds}^s = K_e \omega_r \sin \theta_r \\ e_{qs}^s = -K_e \omega_r \cos \theta_r \end{cases} \tag{3.7}$$

From (3.4), the sliding mode observer is defined as follows:

$$\frac{d}{dt}\hat{i}_s = A \cdot \hat{i}_s + B \cdot (v_s - K_s \cdot \text{sgn}(S)) \tag{3.8}$$

where S is the stator current error and K_s is the gain matrix.

$$K_s = k_s \cdot I \tag{3.9}$$

$$S = [s1 \quad s2]^T = \hat{i}_s - i_s \tag{3.10}$$

The sliding surface is defined upon the stator current errors and is given by

$$e_s = \hat{i}_s - i_s = 0 \tag{3.11}$$

The estimation error dynamic is give by the following equation

$$\dot{S} = \frac{d}{dt}(\hat{i}_s - i_s) = A \cdot (\hat{i}_s - i_s) + B \cdot (E_s - K_s \cdot \text{sgn}(S)) \tag{3.12}$$

The above dynamic function is disturbed by the unknown induced EMF components. In order to guarantee that the sliding surface is achieved, Lyapunov function is used. The Lyapunov function V is chosen as:

$$V = \frac{1}{2}S^T S \tag{3.13}$$

Under the assumption that the rotor speed is constant within one estimated period, the derivative of Lyapunov function becomes:

$$\dot{V} = S^T \cdot \dot{S} = S^T [A \cdot S + B \cdot (E - K_s \cdot \text{sgn}(S))] \tag{3.14}$$

According to Lyapunov's stability theory, $\dot{V}$ must be negative to ensure that the observer is stable. Since parameter matrix A is negative, and the back-EMF component is bounded, the sliding surface is reached after a finite time interval if the sliding gain satisfies

$$K_s > |E|. \tag{3.15}$$

3.3.2 Position Estimation

After sliding surface is reached, comparing (3.4) and (3.8), sliding component $K_s \cdot \text{sgn}(S)$ is the back-EMF with some high frequency harmonics. It may not be suitable to estimate rotor position. So a low pass filter is needed to smooth the back-EMF. It is noted that the electrical time constant is much smaller than the mechanical time constant, especially for high-pole machines. Therefore, it is reasonable to assume that the machine speed is constant over a short time interval. Hence, the dynamic of the back-EMF can be described as [26]:

$$\frac{d}{dt}\begin{bmatrix} e_{qs}^s \\ e_{ds}^s \end{bmatrix} = \begin{bmatrix} 0 & \omega_r \\ -\omega_r & 0 \end{bmatrix}\begin{bmatrix} e_{qs}^s \\ e_{ds}^s \end{bmatrix} \tag{3.16}$$

thus, the filter can be defined as :

$$\frac{d}{dt}\begin{bmatrix} \hat{e}_{qs}^s \\ \hat{e}_{ds}^s \end{bmatrix} = \begin{bmatrix} 0 & \omega_r \\ -\omega_r & 0 \end{bmatrix}\begin{bmatrix} \hat{e}_{qs}^s \\ \hat{e}_{ds}^s \end{bmatrix} - K_f(\begin{bmatrix} \hat{e}_{qs}^s \\ \hat{e}_{ds}^s \end{bmatrix} - K_s \cdot \text{sgn}(S)\begin{bmatrix} 1 \\ 1 \end{bmatrix}) \tag{3.17}$$

The above filter has the structure of a Kalman filter and is expected to have high filtering properties. Cut-off frequency K_f varies according to the rotor speed.

After a smooth back-EMF is obtained, the rotor angle can be calculated by:

$$\theta_r = \tan^{-1} \left| \frac{e_{ds}^s}{e_{qs}^s} \right| \tag{3.18}$$

the sign of e_{ds}^s and e_{qs}^s decides the 4 quadrant of rotor angle. If e_{qs}^s is equal to zero,

$$\theta_r = \begin{cases} \pi/2, & e_{ds}^s > 0 \\ 3\pi/2, & e_{ds}^s < 0 \end{cases} \tag{3.19}$$

The rotor speed can be simply obtained from the derivative of the position.

3.3.3 Hybrid Sliding Mode Observer of PMSM

Though the rotor angle can be obtained through the above sliding mode observer, it has several limitations. i) It cannot be used at the standstill and low speed operation under full load condition; ii) the estimated rotor angle has a phase shift due to the filtering process; iii) the estimated error can be accumulated. Conventionally, the second problem may be solved by rotor angle compensation algorithm though it increases the complexity. However, high frequency harmonics of sliding mode function vary when the rotor speed varies. It is difficult to design one cut-off frequency that works well from low to high speed. If a filter with multiple cut-off frequency is used, on the other hand, it may cause problems for the rotor angle compensation algorithm. Moreover, the first and third problems still remain not to be solved. Thus, the hybrid observer with Hall sensor is introduced in this section to solve these problems.

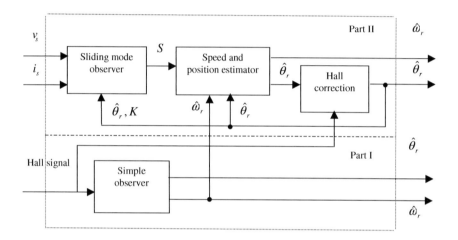

Fig. 3.2 Block diagram of the hybrid observer for PMSM

The structure of hybrid observer is shown in Fig. 3.2. First part is a simple observer that estimates rotor speed and position in the low speed range using Hall sensor signals shown in Fig 3.3. Since the mechanical time constant is large, the rotor speed may be assumed to be constant during every 180/n (where n is the number of Hall sensors) electrical degree interval at low speed range. Under this assumption, motor speed is calculated by:

$$\omega_r = \frac{\pi}{n \cdot \Delta t} \tag{3.20}$$

where Δt is the time interval between two nearest Hall sensor signals. The position can be obtained by integrating speed between these two signals. The initial condition is known from Hall sensors.

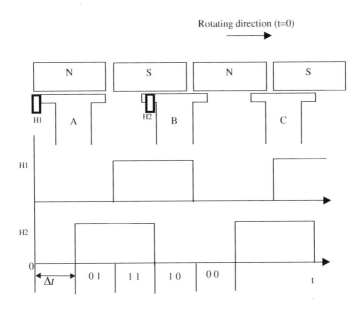

Fig. 3.3 The location and signals of Hall sensors

The second part consists of three blocks: sliding mode current observer described in section 3.3.1, position estimation described in section 3.3.2 and Hall signals correction. The output of the current observer is sliding function with back-EMF information, which is the input of position estimation block. Position estimation block estimates the rotor position using the filtered back-EMF. The varying cut-off frequency ensures the good performance of filter. Finally, the Hall correction block gives more accurate rotor position by eliminating the phase shift and the accumulated error of the rotor position using the real position signals received from Hall sensors.

In the complete control system, initially motor starts and runs to low speed using position and speed information from the developed simple observer. When the motor reaches certain speed, the information from the sliding mode observer is introduced and the simple observer is disabled. The switch from the simple observer to the sliding mode observer is very easy. The sliding mode observer starts working from standstill in parallel to the simple observer. Thus, it is already converged and rotor position is available when motor is running at the low speed. Thus, the switch can occur at any time when motor run at the low speed.

The developed hybrid observer has been verified by simulation and experiments in the entire drive system. The results are given in the following chapter.

3.4 Hybrid Sliding Mode Observer of BLDC Motor

Unlike the typical BLDC drive, in which six rotor positions per cycle are enough to run the motor, the accurate rotor position is required in BLDC motor drive system developed later in order to transfer the line-to-line motor model into the average model. Thus, it is necessary to set forth a position estimator of BLDC motor. Since the BLDC drive system is a multiple-order nonlinear system, sliding mode observer is a good candidate for position and speed estimation. In this section, a hybrid sliding mode observer is developed using typical six rotor positions obtained from Hall sensors.

3.4.1 Sliding Mode Observer

Since the state variables in voltage equation of BLDC are not associated with position in one-to-one relations, it is not suitable to build the sliding mode observer similar to PMSM in the above section. Thus, another structure, which first presented by Husain for SRM [53], is introduced here.

Let us consider a plant of the form

$$\dot{x}_1 = x_2$$
$$\dot{x}_2 = u \tag{3.21}$$

where $X \in \Re^2$ and $u \in \Re$ is the control input.

Let the observer be of the structure

$$\dot{\hat{x}}_1 = \hat{x}_2 - k_1 \operatorname{sgn}(\tilde{x}_1)$$
$$\dot{\hat{x}}_2 = -k_2 \operatorname{sgn}(\tilde{x}_1) + u \tag{3.22}$$
$$\tilde{x} = \hat{x} - x$$

where k_1 and k_2 are positive constants and sgn(x) is the sign function.

Such an observer structure leads to an error dynamics of the form

$$\dot{\tilde{x}}_1 = \tilde{x}_2 - k_1 \operatorname{sgn}(\tilde{x}_1)$$
$$\dot{\tilde{x}}_2 = -k_2 \operatorname{sgn}(\tilde{x}_1) \tag{3.23}$$

If the condition

$$|x_2| \leq k_1$$
$$x_1 = 0 \tag{3.24}$$

is satisfied, then $\tilde{x}_1 = 0$ is attractive and x_2 decay exponentially with a time constant of k_1/k_2. For the detail of proof refer to [53].

3.4.2 Sliding Mode Observer of BLDC motor

Based on the above sliding mode structure, the sliding mode observer for BLDC motor is set forth. The BLDC drive system can be governed by:

$$V_{abcs} = r_s i_{abcs} + \frac{d}{dt}\lambda_{abcs} \tag{3.25}$$

$$\frac{d\theta}{dt} = \omega \tag{3.26}$$

$$\frac{d\omega}{dt} = -\frac{B_m}{J}\omega + \frac{P}{2J}(T_e - T_L) \tag{3.27}$$

where $f_{abcs} = [f_{as} \quad f_{bs} \quad f_{cs}]^T$ are the phase variables of a BLDC motor. J is the combined inertia of the rotor and load. P is the pole number and B_m is the damping coefficient. In the above equations, terminal voltages and currents are the measured variables and ω and θ are not available for measurement, which will be estimated. Based on these equations, a sliding mode observer of BLDC similar to (3.22) may be defined as:

$$\dot{\hat{\theta}} = \hat{\omega} - K_\theta \, \text{sgn}(e_f)$$
$$\dot{\hat{\omega}} = -K_\omega \, \text{sgn}(e_f) \qquad (3.28)$$

The error function is written as:

$$e_\theta = \hat{\theta} - \theta$$
$$e_\omega = \hat{\omega} - \omega \qquad (3.29)$$

Then, the error dynamics is given by:

$$\frac{d}{dt}e_\theta = \hat{\omega}(t) + K_\theta \, \text{sgn}(e_f) - \omega(t) = e_\omega - K_\theta \, \text{sgn}(e_f) \qquad (3.30)$$

$$\frac{d}{dt}e_\omega = -K_\omega \, \text{sgn}(e_f) + \frac{B_m}{J}\omega - \frac{P}{2J}(T_e - T_L) \qquad (3.31)$$

If K_ω can be selected to be large enough such that the last two terms in (3.31) may be neglected, (3.31) can be simplified as

$$\frac{d}{dt}e_\omega = -K_\omega \, \text{sgn}(e_f) \qquad (3.32)$$

According to the above section, this sliding mode observer will be converged in a finite time governed by (3.33) if the sign of the function e_f is opposite to that of the function e_θ:

$$\frac{de_\theta}{dt} = 0$$
$$\frac{de_\omega}{dt} = -\frac{K_\omega}{K_\theta} e_w \quad (3.33)$$

From the above equation, it is clear that e_ω will decay exponentially after the sliding surface $e_\theta = 0$ is reached. Therefore, the objective now becomes to derive the function of e_f.

In general, this function e_f should include a variable dependent on the estimated position and a measurable machine variable. Though the back-EMF includes the rotor position information, it cannot be directly measured and is also not suitable as the correction variable since it includes a series of harmonics, which is related to the rotor position. In order to derive a valid signal from back-EMF, let us look at the motor voltage equation. It is Assumed that harmonics of variables higher than 7 is neglected, which is reasonable since their amplitudes are very small. If three voltage equations are added together, the left side is the 3rd voltage components and the right side are the 3rd resistance drop and the 3rd harmonic component of the back-EMF. If the resistance drop is neglected above a certain speed, the 3rd harmonic back-EMF, e_3, is equal to the 3rd voltage component which is the measurable variable. The estimated one, $\hat{e}_3$, can be calculated from estimated position since the shape of the back-EMF is known in BLDC motor. Hence, the function e_f is defined as:

$$e_f = e_3 - \hat{e}_3 \quad when 0 \le \theta \le 60, 120 \le \theta \le 180, 240 \le \theta \le 300$$
$$e_f = -(e_3 - \hat{e}_3) \quad when 60 \le \theta \le 120, 180 \le \theta \le 240, 300 \le \theta \le 360$$
(3.34)

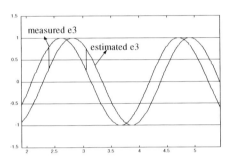

if $e_\theta > 0$

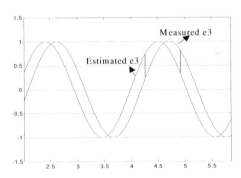

if $e_\theta < 0$

Fig. 3.4 The relation between e_θ and $e_f = e_3 - \hat{e}_3$

Every 60-degree interval can be determined by Hall sensor signals. Fig.3.4 shows the relation between e_f and e_θ. It can be seen that the sliding mode condition is satisfied.

3.4.3 Hybrid Sliding Mode Observer of BLDC Motor

The above observer also suffers from the limitation III and I described in section 3.3.3. In order to solve these problems, the hybrid sliding mode observer used in section 3.3.3 is employed. The simple observer is used at the starting and during low speed operation, the sliding mode observer start to work from medium speed until to high speed and Hall correction part also eliminates the accumulated error and correct the estimated position. The structure of hybrid observer is shown in Fig.3.5. The detail description is not given since it is the same as the one explained in section 3.3.3. On the other hand, for BLDC motor, there is another choice for the starting and low speed operation. Instead of using simple observer with vector control algorithm, six-step current control can be used at the starting and during low speed operation. It results in less torque but with less torque ripple.

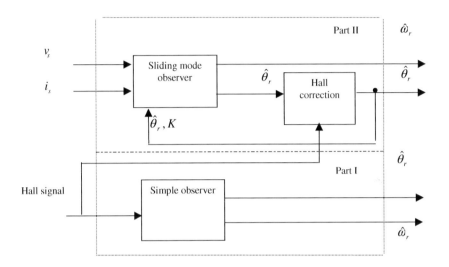

Fig. 3.5 Block diagram of hybrid sliding mode observer of BLDC motor

CHAPTER IV

SURFACE MOUNT PERMANENT MAGNET SPINDLE MOTOR FULL-SPEED OPERATION USING HYBRID SLIDING MODE OBSERVER

4.1 Introduction

As described in Chapter I, Permanent Magnet Synchronous Motor (PMSM) is defined as PM machine with sinusoidal back-EMF. According to the results of motor analysis in Chapter II, the 36-slots/48 poles spindle motor (36/48 spindle motor) matches this definition. Thus, the aim of this chapter is to model the entire spindle motor drive system using the knowledge of PMSM such that the performance of the system can be predicted even before it is built.

The chapter is organized as follow: Section 4.2 first presents the mathematical model of 36/48 spindle motor in abc stationary frame and then d-q model in rotor reference is developed using the new transformation. Based on this d-q model, in section 4.3, the normalized d-q model is first introduced; Secondly the analysis of entire drive system is explained and the optimal control strategies are developed. The adaptive speed control is displayed and especially the flux weakening control is discussed in detail in Section 4.4. Finally, the structure of the entire drive system is verified by the simulation results in section 4.5.

4.2 The Mathematics Model of 36/48 Spindle Motor

4.2.1 The Mathematics Model of 36/48 Spindle Motor in the Stationary Frame

Since the 36/48 spindle motor repeats its configuration and performance every 3-slot 4-pole, it is possible to set forth the mathematical model based on the analysis of 3-slot 4-pole section. Fig.4.1 depicts this part, which has three wye-connected concentrated stator windings (3 slots), and 4-pole, permanent magnet rotor. According to the analysis results in Chapter II, these stator windings are identical windings displaced from each other by 10 mechanical degree, which is corresponding to 240 electrical degree, and each has the resistance r_s. The back-EMF due to the permanent magnet field is sinusoidal wave and repeats itself twice. If one slot corresponds to one phase in the conventional synchronous machine, this section can be equivalent to 3-phase 4 pole synchronous machine in the view of its behavior. Hence, the voltage and torque equation of 36/48 spindle motor can be established by employing the knowledge of synchronous machine. Notice that the electrical degree (24 times mechanical degree) is used in the following analysis.

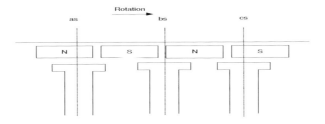

Fig .4.1 3-slot 4-poles part of 36/48 motor

The voltage equation in the abc stationary frame are

$$V_{abcs} = \Lambda_s i_{abcs} + \frac{d}{dt}\lambda_{abcs} \qquad (4.1)$$

where

$$f_{abcs} = [f_{as} \quad f_{bs} \quad f_{cs}]^T \qquad (4.2)$$

$$\Lambda_s = diag[r_s \quad r_s \quad r_s] \qquad (4.3)$$

The flux linkage equation can be expressed as:

$$\lambda_{abcs} = L_s i_{abcs} + \lambda'_m \begin{bmatrix} \sin\vartheta_r \\ \sin(\vartheta_r + \frac{2\pi}{3}) \\ \sin(\vartheta_r - \frac{2\pi}{3}) \end{bmatrix} \qquad (4.4)$$

where λ_m' denotes the amplitude of the flux linkages established by the permanent magnet as viewed from the stator phase windings. Since the 36/48 spindle motor is surface mount motor, the stator self inductance matrix, L_s, is given as:

$$L_s = \begin{bmatrix} L_{ls} + L_{ms} & -\frac{1}{2}L_{ms} & -\frac{1}{2}L_{ms} \\ -\frac{1}{2}L_{ms} & L_{ls} + L_{ms} & -\frac{1}{2}L_{ms} \\ -\frac{1}{2}L_{ms} & -\frac{1}{2}L_{ms} & L_{ls} + L_{ms} \end{bmatrix} \qquad (4.5)$$

The electromagnetic torque may be written as [57]:

$$T_e = \frac{P}{2}\lambda_m'[(i_{as} - \frac{1}{2}i_{bs} - \frac{1}{2}i_{cs})\cos\theta_r - \frac{\sqrt{3}}{2}(i_{bs} - i_{cs})\sin\vartheta_r] + T_{cog}(\theta_r) \quad (4.6)$$

in (4.6), $T_{cog}(\theta_r)$ represents the cogging torque.

The torque and speed is related as

$$J\frac{d}{dt}\omega_{rm} = \frac{P}{2}(T_e - T_L) - B_m\omega_{rm} \quad (4.7)$$

where J is the rotational inertia, B_m approximates the mechanical damping due to friction and T_L is the load torque.

4.2.2 The Mathematical Model of 36/48 Spindle Motor in Rotor Reference Frame

The above voltage and torque equation can be expressed in the rotor reference frame to transfer the time-varying variable to constant value in steady state. Since the stator has two poles and the rotor has four poles, the transformation of 3-phase variables in the stationary frame to the rotor reference frame is different from the conventional park transformation. The new transform is introduced and defined as:

$$f_{qd0r} = K_r f_{abcs} \quad (4.8)$$

where

$$K_r = \begin{bmatrix} \cos\theta_r & \cos(\theta_r + \frac{2\pi}{3}) & \cos(\theta_r - \frac{2\pi}{3}) \\ \sin\theta_r & \sin(\theta_r + \frac{2\pi}{3}) & \sin(\theta_r - \frac{2\pi}{3}) \\ \frac{1}{2} & \frac{1}{2} & \frac{1}{2} \end{bmatrix} \quad (4.9)$$

Based on the knowledge of Chapter II, in order to generate the positive torque the applied stator voltages are:

$$\begin{cases} V_{as} = \sqrt{2}V_s \cos\theta_{ev} \\ V_{bs} = \sqrt{2}V_s \cos(\theta_{ev} + \frac{2\pi}{3}) \\ V_{cs} = \sqrt{2}V_s \cos(\theta_{ev} - \frac{2\pi}{3}) \end{cases} \quad (4.10)$$

Applying (4.9) to (4.1), (4.4) and (4.10) yields:

$$v_{qs}^r = r_s i_{qs}^r + \omega_r \lambda_{ds}^r + \frac{d}{dt}\lambda_{qs}^r \quad (4.11)$$

$$v_{ds}^r = r_s i_{ds}^r - \omega_r \lambda_{qs}^r + \frac{d}{dt}\lambda_{ds}^r \quad (4.12)$$

$$\lambda_{qs}^r = L_s i_{qs}^r \quad (4.13)$$

$$\lambda_{ds}^r = L_s i_{ds}^r + \lambda_m^{`r} \quad (4.14)$$

where $L_s = L_{ls} + L_{ms}$

The electromagnetic torque can be written as:

$$T_e = \frac{3}{2}\frac{P}{2}\lambda_m^{`r} i_{qs}^r \quad (4.15)$$

It can be seen that (4.11)-(4.12) is the same as the conventional PMSM d-q model and can be used in the following analysis.

4.3 Analysis of 36/48 Spindle Motor Drive System

4.3.1 Analysis of 36/48 Spindle Motor Drive System

In order to reduce the parameters, the per-unit system is introduced. The base values are chosen to give one per unit stator current, voltage and flux linkages at one per unit speed when rated currents give maximum torque. Also, the phase resistance is neglected because the motor is used comparatively at high-speed range and the resistance drop becomes very small. Since the PM motor drive system is inverter-fed system, the inverter capability should be taken into account, Considering the inverter capacity, the armature current I_a and the terminal V_a are limited as follows:

$$I_a \leq I_{lim} \tag{4.16}$$

$$V_a \leq V_{lim} \tag{4.17}$$

The current limit, I_{lim}, is decided by the continuous armature current rating and available output current of the inverter. The voltage-limit, V_{lim}, is decided by the available maximum output voltage of the inverter.

If the current limit and voltage limit are selected as their rated values in defined per-unit system, the base values may be given as:

Voltage base: $V_{base} = V_{lim}$ \hfill (4.18)

Current base: $I_{base} = I_{lim}$ \hfill (4.19)

Flux linkage base: $\lambda_{base} = \sqrt{(\lambda_m^{'r} + L_s i_{ds}^r)^2 + (L_s i_{qs}^r)^2}$ (4.20)

Angular frequency base: $\omega_{base} = \dfrac{V_{base}}{\lambda_{base}}$ (4.21)

Inductance base: $L_{base} = \dfrac{\lambda_{base}}{I_{base}}$ (4.22)

Torque base: $T_{base} = \dfrac{3 \cdot V_{base} \cdot I_{base}}{2 \cdot (pole \quad pairs) \cdot \omega_{base}}$ (4.23)

Using the defined base value, the normalized d-q model in steady state is given by:

$V_{dn} = -\omega_{rn} L_{sn} I_{qn}$ (4.24)

$V_{qn} = \omega_{rn} L_{sn} I_{dn} + \omega_{rn} \lambda_{mn}$ (4.25)

$T_{en} = \lambda_{mn} I_{qn}$ (4.26)

$P_n = \omega_{rn} \lambda_{mn} I_{qn}$ (4.27)

Note the use of the subscript "n" to indicate normalized parameters.

Since the motors are normally current-controlled motor, a given operating point can be represented by its location in the (I_d I_q) plane. Hence, the circle diagram [38] is employed to visualize voltage and current of the operating point in steady state and develop the optimal control strategies for spindle motor drive system.

From normalized voltage equation, the voltage limit circle is given by:

$$(\lambda_{mn} + L_{sn} i_{dn})^2 + (L_{sn} i_{qn})^2 = \left(\frac{1}{\omega_{rn}}\right)^2 \quad (4.28)$$

The current limit circle is given as:

$$(i_{dn})^2 + (i_{qn})^2 = 1 \quad (4.29)$$

From (4.27), (4,28) and (4.29), the maximum output power trajectory is given by:

$$I_{dn} = -\frac{\lambda_{mn}}{L_{sn}} \quad (4.30)$$

Then, the overall drive system has three operation modes, which is defined in [40] and illustrated in Fig .4.2 and Fig. 4.3.

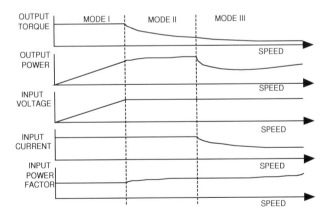

Fig. 4.2 The ideal operation of PM spindle motor

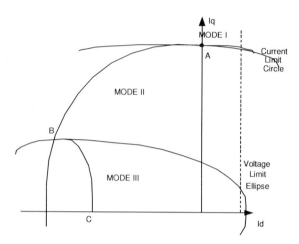

Fig. 4.3 Optimal control strategies for PMSM

- Mode I: current limited region (constant torque region).

This is the region from zero speed to the base speed where maximum torque is obtained by operating at rated current. For the proposed spindle motor, which is surface mount PM motor, this corresponds to point A in Fig. 4.3. In that point, the d-axis current is set to zero and q-axis current is equal to the rated current. The maximum torque accelerates the motor until the terminal voltage reaches its limit value V_{lim} at $\omega = \omega_{base}$ (angular frequency base).

- Mode II: current-and-voltage-limited region.

Here the motor is operated at rated current with the minimum current angle required to give rated terminal voltage. That is, at the intersection of the voltage and current-limit loci (line AB). This intersection point gives the maximum output torque under current and voltage limit. At the same time, it also provides the maximum output power. During this region, the current and voltage keep the same amplitude. The motor speed is extended until point B ($\omega = \omega_2$) because the d-axis current is set to negative value and the flux due to permanent magnet rotor is weakened.

$$\begin{cases} \omega_2 = \dfrac{1}{\sqrt{L_{sn}^2 - \lambda_{mn}^2}} & \text{if } \lambda_{mn} \leq L_{sn} \\ \omega_2 = \dfrac{1}{\lambda_{mn} - L_{sn}} & \text{if } \lambda_{mn} \geq L_n \end{cases}$$

- Mode III: voltage-limited region.

If $L_{sn} > \lambda_{mn}$, the motor can come to mode III after speed is over the certain point. In mode III, the drive operates at the point where the maximum output trajectory intersects with the voltage-limit ellipse (line BC). During this region, the voltage keeps the same aptitude and the current decreases. The motor also gives the maximum output torque under current and voltage limit and the speed can extend to infinite. Since $L_{sn} \leq \lambda_{mn}$ for the proposed 36/48 spindle motor, the maximum output trajectory is outside of current-limit circle and Mode III does not exist. Output power become zero at point D ($I_q = 0, I_d = -1$ and $\omega = \omega_3 = \dfrac{1}{\lambda_{mn} - L_{sn}}$).

4.3.2 Optimal Current Control Strategies

Based on the above analysis, optimal current control strategies, which produce the maximum output power in entire speed range under current and voltage limitation, are developed. Corresponding to three motor operating modes, the strategies consist of three controlled current regions defined as follows:

- Region I: ($\omega \leq \omega_{base}$)

The motor speeds up from zero to base speed. The d-axis current vector is fixed at the point A (Mode I) and is given by:

$$\begin{cases} i_{ds} = 0 \\ i_{qs} = 1 \end{cases} \tag{4.31}$$

- Region II: ($\omega_{base} \leq \omega \leq \omega_2$)

The motor speed is above the base speed and increases until the ω_2. The current vector point moves from A to B along the intersection of the voltage-limit and current-limit circle (Mode II). By solving (4,28) and (4.29), the d and q axes currents are calculated as:

$$\begin{cases} i_{dn} = \dfrac{(1/\omega_{rn})^2 - (\lambda_{mn})^2 - L_{qn}^2}{2 \cdot \lambda_{mn} \cdot L_{sn}} \\ i_{qn} = \sqrt{1 - [\dfrac{(1/\omega_{rn})^2 - (\lambda_{mn})^2 - L_{qn}^2}{2 \cdot \lambda_{mn} \cdot L_{sn}}]^2} \end{cases} \tag{4.32}$$

- Region III: ($\omega \geq \omega_2$)

The motor runs from ω_2 to infinite speed. The current vector moves from B to C along the maximum output trajectory (Mode III) and is given by:

$$\begin{cases} i_{ds} = -\dfrac{\lambda_{mn}}{L_{sn}} \\ i_{qs} = \dfrac{1}{\omega_{rn} \cdot L_{sn}} \end{cases} \tag{4.33}$$

Note for the proposed 36/48 spindle motor, this region does not exist.

4.4 Adaptive Full Speed Range Control Algorithm

The above section provides the theoretical performance of 36/48 spindle motor. However, in practice, there are a lot of parameters such as the load condition, motor parameters, saturation, iron loss [44], which affect the motor behavior. Hence, it is difficult to achieve the optimal current control. In this section, setting the optimal current control as the best performance, an adaptive full speed range control algorithm is developed to run motor in entire speed range. In this dissertation, notice that full speed range means three regions: starting, from low speed to rated speed control (Maximum Torque Control) and over rated speed control (flux weakening control). These three parts are discussed separately and then adaptive rules are applied for overall speed range in the final analysis.

4.4.1 Starting Algorithm

Unlike the ideal case, starting is a big issue in PMSM control. All sensorless methods of surface mount PMSM fail for full load starting. Since Hall sensor signals are available at standstill, a full-load starting algorithm can be developed based on these signals.

Hall sensor signal is shown in Fig. 3.2. Table 4.1 lists the excited phase during alignment and the boundary of initial position. Since the motor has 48 poles on the rotor, the rotor can be aligned to one phase with acceptable movement (less than 3.75 mechanical degree). After the alignment, it is possible to start the motor with simple observer. Since the simple observer can give the rotor position from standstill, the motor starts with the vector control. In other word, the motor starts with the maximum torque. For low pole motors, if alignment is not allowable, the motor can be started using the simple observer with the initial boundary condition.

Table 4.1 Boundary of position and excited phase vs Hall state

Hall signal (H1 H2)	(0 0)	(0 1)	(1 0)	(1 1)
excited phase	B (positive)	C (negative)	B (negtive)	C (positive)
Boundary of position	$0<\theta_r<\pi/2$	$\pi/2<\theta_r<\pi$	$3\pi/2<\theta_r<2\pi$	$\pi<\theta_r<3\pi/2$

4.4.2 Maximum Torque Control

This region corresponds to optimal control region I. In order to obtain maximum torque/ampere ratio, i_{ds} is set to zero. The variations of load or speed give the commanded torque. If the commanded torque is larger than motor maximum torque, the motor maximum torque is selected as the commanded torque. The torque control is achieved by adjusting the q-axis current since i_{qs} is proportional to the motor torque.

4.4.3 Flux Weakening Control

Flux weakening control is used when the motor speed is above the rated speed, which corresponds to optimal control II and III.

For PM motor, when motor speed is above the rated speed, the motor torque decreases very quickly since the rotational back-EMF rapidly approaches the terminal voltage. Hence, motor speed can be extended very little. In order to solve this problem, the flux-weakening technique is developed. This technique can be explained by phasor diagram as shown Fig. 4.4.

For surface mount PM motor, phase equations in d-q model can be written as:

$$\overline{\varphi}_{as} = x_{qs}\overline{I}_{as} - j\overline{E}_i$$

$$\overline{V}_{as} = r_s\overline{I}_{as} + j\overline{\varphi}_{as}$$

where

$$x_{qs} = \omega_r L_{qs}$$

$$\overline{f}_{as} = f_{qs} - jf_{ds}$$

$$E_i = \omega_r \lambda_m$$

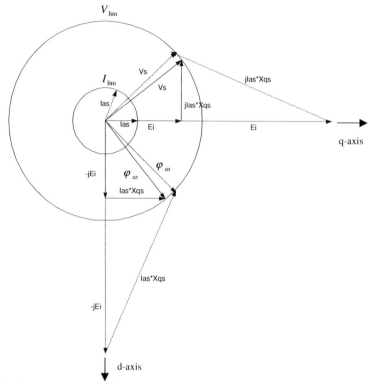

Fig. 4.4 Phasor diagram of PM motor considering the current and voltage limitation

In Fig 4.4, solid lines show the operating point for constant torque, where the motor reaches the maximum speed under maximum motor torque. The dotted lines show the flux weakening operating point, where the motor speed is extended three times, considering the current and voltage limitations. From this Fig, it is clear that the two parameters play the key roles for flux weakening control. The necessary one is that the negative d-axis current should be injected to weaken the field produced by the permanent magnet rotor. Another one is the x_{qs}, which affects the extension of speed range. From Fig.4.5, it can be seen that the extension of speed increases when the $x_{qs}I_{as}/E_i$ increases. On the condition of $E_i \equiv x_{qs}I_{as}$, ideally the output power does not decrease after over rated speed and the motor speed could be extended to infinity. On the

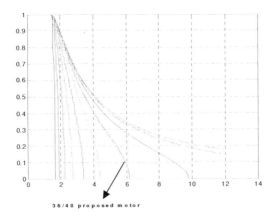

Fig 4.5 Normalized torque vs speed curves

when $x_{qs}I_{as}/E_i$ increases from 0.1 to 1 with 0.1 step

condition of $E_i \geq x_{qs}I_{as}$, the output power decreases until zero as the rotor speed increase to $\omega = \omega_3$. The marked line shows the 36/48 motor condition.

Based on the above concept, the practical flux weakening control can be developed and explained by the following flow chart.

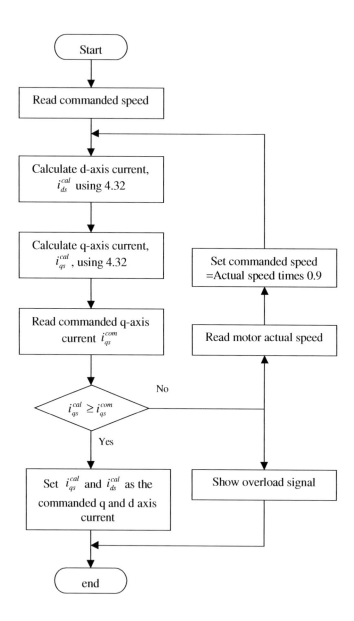

Fig. 4.6a The flow chart of flux weakening control

4.4.4 Adaptive Full Speed Range Control System of PM Spindle Motor

The block diagram of the adaptive full speed range control strategy is shown in Fig.4.6b. The overall system includes two closed-loops, an inner current loop and an outer speed loop. Whenever a reference speed ω_r^* is given, the system automatically compares it with the actual speed ω_r. According to the motor equation of motion, the speed error $\Delta\omega_r$ directly affects the torque profile. Therefore, the output of speed PI regulator is considered as torque reference value T_{ec}^* that is proportional to i_{qsr}^*, where i_{dsr}^* is set to zero during all time. Thus, when the actual speed ω_r or load T_L suddenly changes, i_{dsr}^* and i_{qsr}^* immediately adjust speed and torque. Once the proper adjustment has been accomplished, the motor speed ω_r should follow the given value ω_r^*, and the motor quickly achieves steady state operation. The current loop forces the actual stator

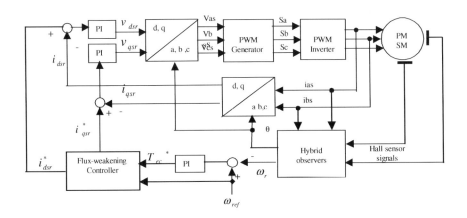

Fig .4.6b Block diagram of PMSM drive system

currents to track the commanded currents. Rotor position can be obtained accurately using the proposed hybrid observer. Since the parameter varies along the long speed range. Adaptive PI controller with respect to motor speed is used in the above system.

4.5 Simulation Analysis

Simulations have been done in Matlab/SIMILINK to verify the developed system. The initial values of state variables in the observer are assumed to be known. Fig.4.7 shows the speed response of PMSM drive system. It is clear that the motor speed can track the commanded speed under full load from zero speed to maximum speed. Fig.4.8 shows actual q axis current in stationary reference frame and estimated current from sliding mode current observer when load varies. Based on this observer, the estimated rotor position is displayed in Fig.4.9. It can be seen that the hybrid sliding mode observer works well. Fig.4.10 and Fig. 4.11 illustrate the motor d and q axis currents in rotor reference frame and the electromagnetic torque and load under the speed command shown in Fig. 4.7. The d-axis current stays at zero during the constant torque control and becomes negative to weaken the field produced by permanent magnet rotor. The q-axis current varies according to the load. All these figures prove the validity of full speed range control.

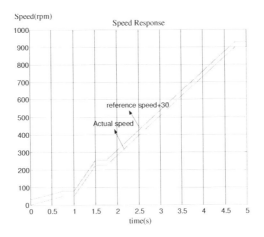

Fig.4.7. Speed response of PMSM

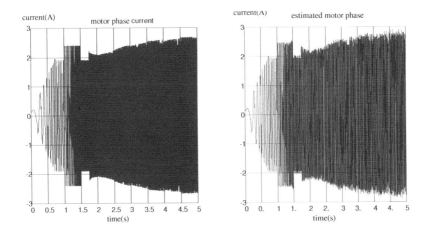

Fig. 4.8 Motor actual and estimated q-axis currents in the stationary frame

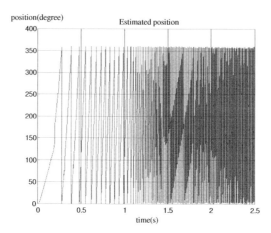

Fig. 4.9 Estimated position from the hybrid observer

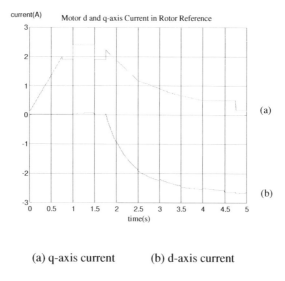

(a) q-axis current   (b) d-axis current

Fig. 4.10 Q and d axes currents in the rotor reference frame

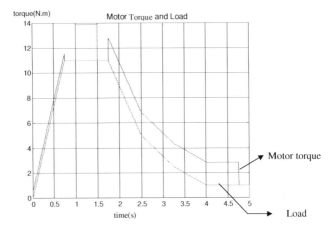

Fig. 4.11 Motor torque and load

CHAPTER V

BLDC MOTOR FULL-SPEED OPERATION USING HYBRID SLIDING MODE OBSERVER

5.1 Introduction

Unlike PMSM drive system which d-q model is employed to achieve constant torque and constant power operation, conventionally, the quasi-square wave (six-step) current control is used to develop the constant torque in BLDC motor drive system as shown in Fig.5.1 and Fig. 5.2. At any given time instant, only two of the switches are conducting. The switches T_1, T_3 and T_5 each carry the positive phase currents and switches T_2, T_4 and T_6 carry the negative phase currents. The DC link current flows through one switch from the upper group and returns through a switch from the lower group. Each switch conducts for 120 electrical degree. So there is a phase commutation at every 60-degree interval. The commutation is defined as the process in which one motor phase stops conducting and another one starts conducting. At all the time the magnitude of the motor phase current is the same as that of the DC link current, so only one current sensor is needed to get information about all the three motor phase currents. Controlling the switching times of the individual switch controls the magnitude of the phase current. The three Hall effect sensors provide position signals to synchronize phase current with the corresponding back-EMF in order to generate the constant torque. The electromagnetic torque generated by the motor is given by,

$$T_m = \frac{(i_a E_a + i_b E_b + i_c E_c)}{\omega_m} \tag{5.1}$$

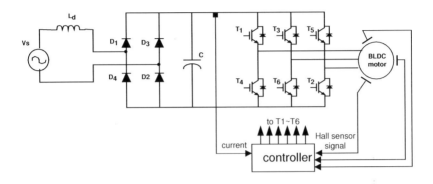

Fig. 5.1 BLDC motor drive system

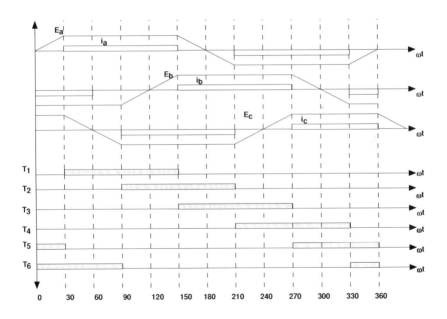

Fig. 5.2 Constant torque control strategy

Where,

E_a, E_b, E_c are motor phase back-emfs in volts,

i_a, i_b, i_c are motor phase currents in ampere and

ω_m is mechanical angular velocity in rad/sec

The ideal waveforms show that torque produced is ripple free and is directly proportional to the motor phase current as in the case of the separately excited DC motor.

In case of the constant power operation, since the conventional d-q model cannot be directly applied to the BLDC motor drive it is difficult for the BLDC motor to achieve it. Hence, if the motor speed is above the rated speed, typically advanced angle control is introduced in the flux-weakening operation. By increasing the lead angle between the phase current and phase back-emf, the flux produced by the permanent magnet can be weakened so that the motor speed is extended without increasing DC bus voltage. However, this comes at the cost of large torque ripple especially for angles greater than 30-electrical degree.

In summary, though the conventional torque or speed control technologies of BLDC motor have proved very useful in many cases, there exists the following drawbacks:

- The capabilities of inverter and motor have not been totally exploited due to 120 degree conducting for each phase. Motor delivers less torque than it could be.

- It is difficult to achieve quantitative speed or torque control in BLDC drive system as compared with PMSM drive in flux-weakening region. In other words, either at a given torque or a given speed, advance-angle to be applied is not exactly known. It limits the high-speed operation of BLDC motor, especially for two major sectors of consumer market, traction and residential markets such as the electric vehicles and washing machines.

This chapter proposes a method that is similar to field oriented control in sinusoidal PM motor, in which at every instant of time all of the three phases conduct, and every switch conducts for 180 degree. Thus, it increases the motor torque during constant torque operation and also exhibits the same controllability as the sinewave PM motor during flux weakening operation.

5.2 Modeling and Analysis of BLDC Machine in Multiple Reference Frame (MRF)

5.2.1 BLDC Line-Line Model in Stationary Frame

In general, the stator winding of BLDC motor is connected in Y configuration to eliminate the circulating third harmonic current. This section sets forth the BLDC model based on Y-connection. The motor structure is assumed to be symmetric surface mount motor. The magnet saturation and demagnetization are neglected. Furthermore, The back EMF is assumed ideal as shown in Figure 5.3. The Fourier series of this waveform can be written:

$$e(t) = \frac{4}{\alpha\pi} \cdot E_m \sum_{k=1,3,5,7,\ldots} \frac{1}{k^2} \cdot \sin k\alpha \cdot \sin kw_r t \tag{5.2}$$

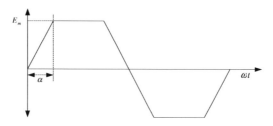

Fig. 5.3 The ideal back EMF

Figure 5.4 depicts a cross section of the simplified three-phase surface mounted BLDC motor considered inhere. The stator windings, as-as', bs-bs' and cs-cs' are shown as lumped windings for simplicity but are actually distributed about the stator and rotor has two poles. Mechanical rotor speed and position are denoted as ω_{rm} and θ_{rm}, respectively. Electrical rotor speed and position, ω_r and θ_r, are defined as P/2 times the corresponding mechanical quantities, where P is the number of poles.

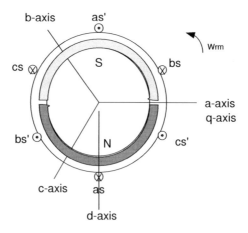

Fig. 5.4 The cross section of BLDC motor

Although third harmonic current becomes zero in Y connection, third harmonic in phase voltage and phase back EMF still exists. In order to derive the uniform motor model, line-to-line variable model is used to eliminate the third variables.

Based on the above description, the stator line-line voltage equation of BLDC can be written as

$$V_{abc} = r_{sl} i_{abc} + \frac{d}{dt} \lambda_{abc} \tag{5.3}$$

where $f_{abc} = [f_{ab} \; f_{bc} \; f_{ca}]^T$ represents the line-line current, voltage and flux linkage vectors, respectively. r_{sl} is the stator resistance of line-to-line model and equal to stator phase resistance.

Considering the shape of back EMF as shown in Fig.5.3 and assuming $\lambda_{ab} = 0$ when $\theta_r = 0$, then the stator line-to-line flux linkage equation can be expressed as:

$$\lambda_{abc} = L_s i_{abc} + \lambda_m' \sum_{n=1,5,7}^{\infty} K_{en} \begin{bmatrix} \sin n\vartheta_r \\ \sin n(\vartheta_r - \frac{2\pi}{3}) \\ \sin n(\vartheta_r + \frac{2\pi}{3}) \end{bmatrix} \tag{5.4}$$

where λ_m' denotes the peak strength of the fundamental component of magnet flux linkage and is written as:

$$\lambda_m' = \frac{8 \cdot E_m}{\alpha \cdot \pi \cdot \omega_r} \sin \alpha \cdot \sin \frac{\pi}{3} \tag{5.5}$$

The stator self inductance matrix of line-to-line model, L_s, may be given as:

$$L_s = \begin{bmatrix} L_{ls} + L_{ms} & -\frac{1}{2}L_{ms} & -\frac{1}{2}L_{ms} \\ -\frac{1}{2}L_{ms} & L_{ls} + L_{ms} & -\frac{1}{2}L_{ms} \\ -\frac{1}{2}L_{ms} & -\frac{1}{2}L_{ms} & L_{ls} + L_{ms} \end{bmatrix} \quad (5.6)$$

L_{ls} and L_{ms} denote the phase leakage and mutual inductances, respectively. The coefficient K_n represents the magnitude of the nth harmonic of magnet flux relative to the fundamental component and is given as:

$$K_{en} = \frac{1}{n^3} \sin(n\alpha) \cdot \sin(n \cdot \frac{\pi}{3}), \text{ n=1,5,7,11,...} \quad (5.7)$$

The electromagnetic torque may be derived from coenergy as given by:

$$T_e = \frac{P}{4}\lambda'_m \sum_{n=1,5,7}^{\infty} K'_{en} \begin{bmatrix} i_{ab} \\ i_{bc} \\ i_{ca} \end{bmatrix}^T \begin{bmatrix} \cos n\theta_r \\ \cos n(\theta_r - \frac{2\pi}{3}) \\ \cos n(\theta_r + \frac{2\pi}{3}) \end{bmatrix} + T_{cog}(\theta_r) \quad (5.8)$$

in (5.8)

$$K'_{en} = nK_{en} \quad \text{n = 1,5,7,...} \quad (5.9)$$

and $T_{cog}(\theta_r)$ represents the cogging torque.

The combined mechanical dynamics of the machine and load are represented by

$$J\frac{d}{dt}\omega_{rm} = T_e - T_L - B_m\omega_{rm} \tag{5.10}$$

where J is the rotational inertia, B_m approximates the mechanical damping due to friction and T_L is the load torque.

Equation 5.3~5.10 govern the Line-to-line BLDC motor drive system and are the basis of the following analysis.

5.2.2 BLDC Model in the Rotor Multiple Reference Frame(MRF)

Though the line-to-line model is mathematically and physically valid, it is not suited to design the control system since the currents have not been decoupled and torque varies with the rotor position. In PMSM machine with sinusoidal back-EMF, considerable simplification has been obtained by transforming the abc model in stator reference frame to dq model in the rotor reference frame. However, if the same transformation is applied to line-to-line model, the current, flux and torque are still time-varying and not constant at steady state [55]. In order to solve these problems, multiple reference transformation [55](Note that the transformation presented in [55] is wrong) and average model [56] are introduced.

Multiple reference frame transformation is defined as:

$$f_{qd}^{xr} = K_s^{xr} f_{abc} \tag{5.11}$$

where

$$K_s^{xr} = \frac{2}{3}\begin{bmatrix} \cos x\theta_r & \cos x(\theta_r - \frac{2}{3}\pi) & \cos x(\theta_r + \frac{2}{3}\pi) \\ \sin x\theta_r & \sin x(\theta_r - \frac{2}{3}\pi) & \sin x(\theta_r + \frac{2}{3}\pi) \\ \frac{1}{2} & \frac{1}{2} & \frac{1}{2} \end{bmatrix} \quad (5.12)$$

It represents a transformation from the stator reference frame into the 'xr' reference frame rotating at 'x' times of electrical speed.

Applying (5.11) to (5.3) yields the voltage equation in MRF as:

$$v_{qs}^{xr} = r_{sl} i_{qs}^{xr} + x\omega_r \lambda_{ds}^{xr} + \frac{d}{dt}\lambda_{qs}^{xr} \quad (5.13)$$

$$v_{ds}^{xr} = r_{sl} i_{ds}^{xr} - x\omega_r \lambda_{qs}^{xr} + \frac{d}{dt}\lambda_{ds}^{xr} \quad (5.14)$$

Following same procedure, the flux linkage in MRF can be given by:

$$\lambda_{qs}^{xr} = L_{ss} i_{qs}^{xr} + \lambda_m' \sum_{n=1}^{\infty}[K_{6n-1}\sin((x-1+6n)\theta_r) - K_{6n+1}\sin((x-1-6n)\theta_r)] - \lambda_m'\sin((x-1)\theta_r) \quad (5.15)$$

$$\lambda_{ds}^{xr} = L_{ss} i_{ds}^{xr} + \lambda_m' \sum_{n=1}^{\infty}[-K_{6n-1}\cos((x-1+6n)\theta_r) + K_{6n+1}\cos((x-1-6n)\theta_r)] + \lambda_m'\cos((x-1)\theta_r) \quad (5.16)$$

where $L_{ss} = L_{ls} + \frac{3}{2}L_{ms}$

The torque equation in MRF is written as:

$$T_e = \frac{P}{4}\lambda_m'[i_{qs}^r + \sum_{n=1}^{\infty}(K_{6n-1}' \cdot i_{qs}^{(1-6n)r} + K_{6n+1}' \cdot i_{qs}^{(6n+1)r})] + T_{cog}(\theta_r) \quad (5.17)$$

Note that the voltage equations have the same form as the conventional dq model used in sinusoidal PM motor and torque equation is only written in terms of magnet strength, λ_m', and q-axis currents in the different reference frame. However, the flux linkages that are used as state variables still contain position dependent term. Therefore, an average procedure defined in [56] is employed to make the state variable independent and to yield variables, which are constant at steady state.

5.2.3 Average Model of BLDC Motor in MRF

This procedure is based on the observation that the voltage, current, and flux linkages will vary as a function of $a6\theta_r$ (where a is the set of positive integers) in the reference frames of interest. Hence, the average model is obtained by simply replacing the variable in MRF model with their average values taken over the period corresponding to a $2\pi/6$ increment of electrical rotor position.

The entire average model is represented by:

$$\bar{v}_{qs}^{xr} = r_s \bar{i}_{qs}^{xr} + x\omega_r \bar{\lambda}_{ds}^{xr} + \frac{d}{dt}\bar{\lambda}_{qs}^{xr} \tag{5.18}$$

$$\bar{v}_{ds}^{xr} = r_s \bar{i}_{ds}^{xr} - x\omega_r \bar{\lambda}_{dqs}^{xr} + \frac{d}{dt}\bar{\lambda}_{ds}^{xr} \tag{5.19}$$

$$\bar{\lambda}_{qs}^{xr} = L_{ss}\bar{i}_{qs}^{xr} \tag{5.20}$$

$$\bar{\lambda}_{ds}^{xr} = L_{ss}\bar{i}_{ds}^{xr} + \frac{|x|}{x}K_{e|x|}\lambda_m' \tag{5.21}$$

$$\overline{T}_e = \frac{3P}{4}\lambda'_m[\overline{i}^r_{qs} + \sum_{x=1}^{\infty}(K'_{6n-1}\overline{i}^{(1-6n)r}_{qs} + K'_{6n+1}\overline{i}^{(1+6n)r}_{qs})] \qquad (5.22)$$

Note that 5.18 - 5.22 should be written for each of the reference frame considered.

The average model performs the desired characteristics. These include the absence of rotor position terms and voltage, current, and flux linkage which are constant at the steady state. However, it is unnecessary to use all harmonics component to design the control system, Table 5.1 shows the amplitudes of nominated harmonics back-EMF which become very small when n is larger than 7 and the first, fifth, and seven harmonics components have 99% of its power. Thus, in the following analysis it is reasonable to limit x to be equal to 1, -5, 7 and n is equal to 1.

Table 5.1 The normalized amplitude of back- EMF harmonics components

	α=30	α=25	α=20	α=15	α=10	α=5	α=0
e_{ab1}	1	1.01428	1.02606	1.0353	1.04189	1.04587	1.0472
e_{ab5}	-0.04	-0.07864	-0.1182	-0.1545	-0.01838	-0.2029	-0.2094
e_{ab7}	-0.0204	0.0427	0.03935	0.07885	0.1151	0.1405	0.1496
e_{ab11}	0.0083	0.0195	0.0155	-0.0083	-0.0447	-0.0777	0.0952

It is still inconvenient to develop speed or torque control algorithm based on torque equation of average model because torque is related to the three current components, the fundamental, fifth and seventh harmonic currents. Fig. 5.5 and Fig. 5.6 show the amplitude of the sum of fifth and seventh and the fundamental components of the electromechanical torque. It can be seen that the sum of fifth and the seventh harmonics torque are less than 1% of the total torque. Thus it is reasonable to neglect harmonics torque in average model and express the torque equation as follows:

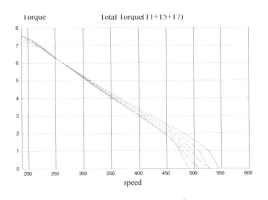

Fig. 5.5 The total torque when the back EMF has the different α angles

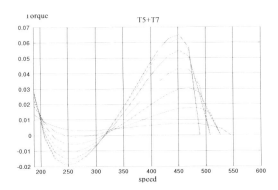

Fig.5. 6 The sum of fifth and seventh harmonics torque when the back EMF has the different α angles

$$\bar{T}_e = \frac{3P}{4} \lambda'_m \bar{i}^r_{qs} \tag{5.23}$$

Though many simplifications are inherent in the torque equation, it is an excellent approximation of real torque equation as long as the system inertia is not small like in washing machine and electric vehicle.

In order to reconstruct the line variables after MRF variables have been solved, the inverse multiple reference transformation is necessary and defined as:

$$\begin{cases} f_{ab} = \sum_{n \in N} [\bar{f}^{nr}_{qs} \cos(n\theta_r) + \bar{f}^{nr}_{ds} \sin(n\theta_r)] \\ f_{bc} = \sum_{n \in N} [\bar{f}^{nr}_{qs} \cos n(\theta_r - 2\pi/3) + \bar{f}^{nr}_{ds} \sin n(\theta_r - 2\pi/3)] \\ f_{ca} = \sum_{n \in N} [\bar{f}^{nr}_{qs} \cos n(\theta + 2\pi/3) + \bar{f}^{nr}_{ds} \sin n(\theta_r + 2\pi/3)] \end{cases} \tag{5.24}$$

At this point, in summary, since (5.23) has the same from as the one of sinusoidal PM in d-q reference frame, the currents and flux linkages are decoupleed in average voltage and flux equations (5.18-5.21), and line-to-line variables can be calculated from (5.24), it is possible to develop control algorithm for BLDC motor using the knowledge that is suitable for sinewave PM motor.

5.3 Full Speed Range Operation Control Algorithm

5.3.1 Constant Torque Control Algorithm

In the average model, there are six components of the currents, the first, the fifth and the seventh harmonics, which should be controlled. The $\bar{i}^r_{qs1}$ can be obtained from torque equation (5.23) of average motor model and $\bar{i}^r_{ds1}$ usually is set to zero in constant torque control in order to achieve maximum ratio of torque/current. However, the fifth and the seventh current harmonics are the free controlled variables and there are many possibilities to select their values. In order to maximize the capability of inverter, the voltage or current limitation of inverter (it is impossible to use both of them as it does in sinewave PM motor) is introduced as a limitation. Since the flux weakening operation should be taken into account in overall control system and motor current cannot be suddenly changed, DC bus voltage limitation is selected as the restriction to obtain the commanded fifth and seventh harmonics currents.

It is known that the maximum phase voltage obtained from the inverter is the half-wave symmetric, which include first, third, fifth and seventh components. The task of the fifth and seventh harmonics currents is to adjust the voltage harmonics such that the phase voltage can almost keep the half–wave symmetric shape in overall control system. The amplitude of voltage varies corresponding to the speed during constant torque control region. After the speed is over the rated speed, the amplitude and shape of voltage keep unchanged and the control strategy of constant power is employed to extend the motor speed.

Based on the above concept, the line-to-line (ab) voltage, which can be calculated from half–wave symmetric three phase voltage, is shown in Figure 5.7. The Fourier series for this wave is given by:

$$f_{ab} = \frac{4}{\pi} \cdot V \sum_{k=1,5,7,..} \frac{1}{k} \cdot [2\sin k(wt + \frac{\pi}{6} - \beta) \cdot \sin k \cdot \frac{\pi}{3}] \qquad (5.25)$$

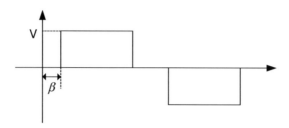

Figure 5.7 line-to line motor voltage

The bc and ca quantities are given by substituting $(\theta_r - \frac{2\pi}{3})$ and $(\theta_r + \frac{2\pi}{3})$ for θ_r into (5.25), respectively.

Applying the MRF transform and average procedure to the three line-to-line voltages yields

$$\overline{V}_{qsx}^r = K_x \cdot \sin x(\frac{\pi}{6} - \beta)$$
$$\overline{V}_{dsx}^r = K_x \cdot \cos x(\frac{\pi}{6} - \beta)$$
(5.26)

where

$$K_x = \frac{9V}{x\pi} \sin \frac{x\pi}{3}$$
(5.27)

and x=1,5, 7 for the fundamental, fifth and seventh harmonics voltages, respectively.

The fundamental average voltage equations can be written as:

$$\overline{V}_{qs1}^r = r_s \overline{i}_{qs1}^r + \omega_r L_d \overline{i}_{ds1}^r + L_q \frac{d}{dt} i_{qs1}^r + \omega_r K_{e1} \lambda_m'$$
(5.28)

$$\bar{V}_{1ds1}^{r} = r_s \bar{i}_{ds1}^{r} - \omega_r L_q \bar{i}_{qs1}^{r} + L_d \frac{d}{dt} i_{ds1}^{r} \tag{5.29}$$

Since the $\bar{i}_{qs1}^{r}$, $\bar{i}_{ds1}^{r}$, ω_r and $\lambda_m^{'}$ are known, the fundamental voltage can be calculated form above voltage equations. Then the angle of β in (5.26) is given:

$$\beta = \frac{\pi}{6} - \tan^{-1}(\frac{\bar{V}_{qs1}^{r}}{\bar{V}_{ds1}^{r}}) \tag{5.30}$$

After β is calculated, the fifth and seventh d-q harmonics voltages can be obtained from (5.26). Based on the voltage equation of average mode, thus, the commanded fifth and seventh harmonics currents can be calculated by:

$$\frac{d}{dt}\bar{i}_{qs5}^{r} = \frac{1}{L_q}(-r_{ls}\bar{i}_{qs5}^{r} - 5\omega_r L_d \bar{i}_{ds5}^{r} + \bar{V}_{qs5}^{r} - 5\omega_r K_{e5}\lambda_m^{'}) \tag{5.31}$$

$$\frac{d}{dt}\bar{i}_{ds5}^{r} = \frac{1}{L_d}(-r_{ls}\bar{i}_{ds5}^{r} + 5\omega_r L_q \bar{i}_{qs5}^{r} + \bar{V}_{ds5}^{r}) \tag{5.32}$$

$$\frac{d}{dt}\bar{i}_{qs7}^{r} = \frac{1}{L_q}(-r_{ls}\bar{i}_{qs7}^{r} - 7\omega_r L_d \bar{i}_{ds7}^{r} + \bar{V}_{qs7}^{r} - 7\omega_r K_{e7}\lambda_m^{'}) \tag{5.33}$$

$$\frac{d}{dt}\bar{i}_{ds7}^{r} = \frac{1}{L_d}(-r_{ls}\bar{i}_{ds7}^{r} + 7\omega_r L_q \bar{i}_{qs7}^{r} + \bar{V}_{ds7}^{r}) \tag{5.34}$$

5.3.2 Constant Power Control Algorithm

It is easy to develop flux weakening control algorithm in terms of the average model after constant torque control is developed in the above section. The developed procedure is the same as PM sinwave motor does. Instead of setting $\bar{i}_{ds1}^{r}$ to be zero,

$\bar{i}_{ds1}^r$ can be calculated from fundamental voltage equation based on the speed. It is assumed that the fundamental phase voltage is equal to maximum voltage in (5.25) and the stator resistance is neglected during flux weakening operation. Hence, $\bar{i}_{ds1}^r$ and $\bar{i}_{qs1}^r$ can be given by:

$$\bar{i}_{ds1}^r = \frac{1}{L_d}[\sqrt{(\frac{4\sqrt{3}}{\pi \cdot \omega}V_{max})^2 - (L_q \bar{i}_{qs1}^r)^2} - \lambda_m'] \tag{5.35}$$

$$\bar{i}_{qs1}^r = \frac{4T_e^*}{3P\lambda_m} \tag{5.36}$$

$$T_e^* = T_{ec}^* \quad if \quad T_{ec}^* < T_{e\,max} \quad , \quad otherwise \quad T_e^* = T_{e\,max} \tag{5.37}$$

$$T_{e\,max} = \frac{3P}{4}\lambda_m'\sqrt{i_{sn1}^2 - (\bar{i}_{ds1}^r)^2} \tag{5.38}$$

where T_{ec}^* is the commanded torque obtained from speed PI and i_{sn1} is the fundamental current limit.

The fifth and seventh harmonics currents can be obtained by following the same algorithm developed in the constant torque control section.

Combining the constant torque and constant power control algorithm, the block diagram of full speed range algorithm is shown in Fig. 5.8.

5.3.3 Starting Algorithm

Three hall sensors provide two choices for starting procedure of BLDC motor. One is the same as the algorithm developed in PMSM chapter. A simple observer provides the position and speed. The constant torque control is employed to start the motor. Another choice is the conventional six-step BLDC control algorithm. The first one gives the larger starting torque and the second one provides the less torque ripple and starting current.

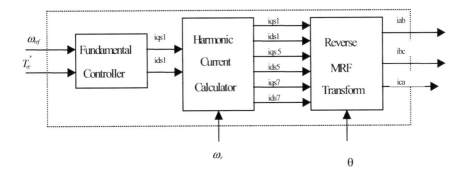

Fig. 5.8 The block diagram of full speed range controller

5.4 BLDC Motor Full Speed Range Control System

The full speed range control strategy block diagram is shown in Fig .5.9. The overall system includes two closed-loops, an inner current loop and an outer speed loop. The speed loop performs the same way as the PMSM does. The variations of speed or torque adjusts the commanded three line-to-line currents using the full-speed range controller as shown in Fig. 5.8. The hybrid sliding mode observer introduced in Charpter III provides the actual rotor speed, ω_r and position, θ .Since no neutral connection is avaible in the proposed BLDC motor, it is necessary to obtain three phase commanded current to control the motor. Hence, three phase commended currents are calculated from three balance line-to-line commanded currents. After commanded currents are calcualted from speed loop, the inner current loop forces actual motor phase currents to track the commanded phase currents by employing two PI controllers. The outputs of PI controllers are the voltages which are applied to motor. Sinsoidal-triangle PWM generator is used to generate the gate signals which control the voltage source inverter. Once the proper adjustment has been accomplished, the motor speed and torque are under control.

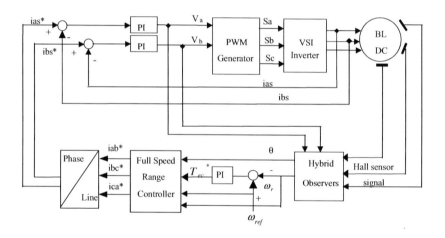

Fig. 5.9 Block diagram of BLDC drive system

5.5 Simulation Results

MATLAB/Simulink simulations have been done to verify the performance of the above system. Figure 5.10 - 5.12 display the constant torque operation. It is observed that the actual speed tracks commanded speed very well and motor starts with full load.

Fugure 5.12 shows the the motor currents which conduct continuous 180 degree in positive and negative values and results in the increase of the developed motor torque compared to the conventional discontinuous current mode in six step current control.

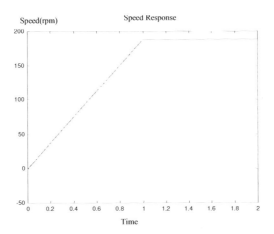

Fig 5.10 Speed response (commanded and actual speed) in constant torque range

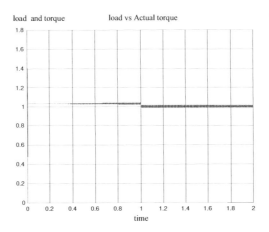

Fig 5.11 Load and motor torque

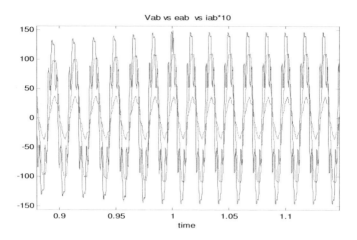

From top to bottom: L-to- L voltage   L-to-L back-EMF

10 times L-to-L current

Fig. 5.12   L-to-L voltage, back-EMF and 10 times the current

Figure 5.13 - 5.16 shows the speed response and sliding mode observer output in the flux weakening operation. Clearly, sliding mode observer performances very well and the estimated position and speed errors are in acceptable range. The constant power operation is illustrated in Figure 5.17 - 5.19. By injecting negative d-axis current, the motor speed is extended even when the back-EMF is much higher than terminal voltage as shown in Figure 5.19. At the same time, q-axis current varies according the load condition as shown in Figure 5.17 and 5.18. All these figures prove that the proposed method shows the same controllability in flux weakening range as the sinusoidal PM motor does.

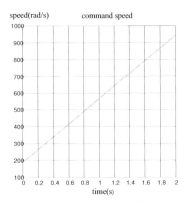
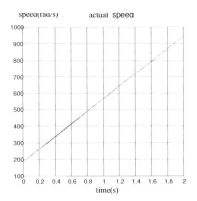

Fig. 5.13 Commanded speed and actual speed in constant power range

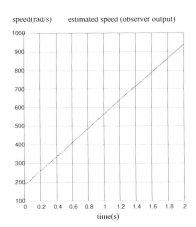
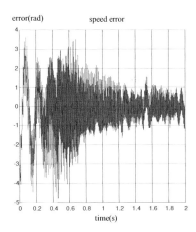

Fig. 5.14 Observer output and estimated speed error

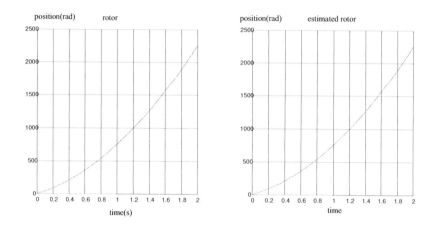

Fig. 5.15 Rotor position and estimated position

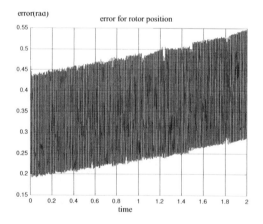

Figure 5.16 Estimated position error

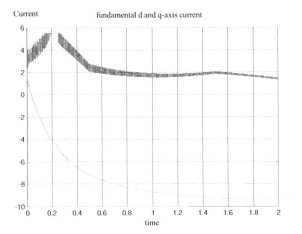

From top to bottom: Q-axis current; D-axis current

Fig. 5.17 Fundamental q- and d-axis current in MRF

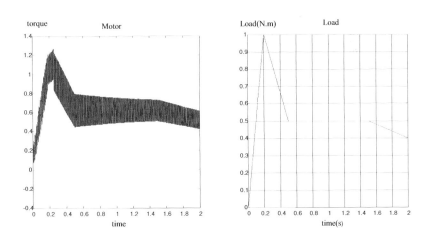

Fig 5.18 Motor torque and load.

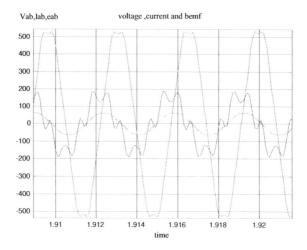

From top to bottom: L-to-L back-EMF   L-to-L voltage

10 times L-to-L current

Fig 5.19 Motor line-to-line voltage, back-EMF, and 10times current

CHAPTER VI

SYSTEM IMPLEMENTATION

6.1 Introduction

Though the simulation results obtained verify the basic operations and performances of the system, the simulation program is based on an ideal model. A variety of practical factors such as saturation, iron loss can influence the performance of the actual motor. Furthermore, the proposed technology will be put into practice for washing machine applications. Hence it is necessary to build a prototype system in the lab to evaluate it practically.

For this purpose, a prototype motor drive system was designed and built in the lab of Motor drive group of the Whirlpool Corporation. An evaluation board developed by Spectrum Digital Inc, (TMS320LF2407 EVM) for Texas Instruments digital signal processor, TMS320LF2407 is used as a controller [58]. TMS320F2407 belongs to the family of fixed-point digital signal processors, designed for motion control applications especially for AC induction, BPM, Switched Reluctance, and Stepper Motor Control [59-60]. The use of such a dedicated controller eliminates a lot of external peripheral circuits and additional software usually needed for motor control using conventional micro-controllers. It has two Event Manager (EV) Modules (A and B) EVA and EVB, where each includes,

- Two 16-bit general-purpose hardware timers,

- Eight 16-Bit Pulse-Width Modulation (PWM) channel,

- Three Capture units,

- Two 10-bit Analog-to-Digital Converters

In the following sections, the system implementation based on TMS320F2407 is explained.

6.2 Experimental Setup

Experiments had been implemented on the 300w, 48 poles, 36 slots PM spindle motor with the concentrated windings as shown in Fig. 6.1. The motor phases are star-connected forming a floating neural connection and two Hall sensors are arranged on the stator. As illustrated in Fig. 2.4, the back-EMF is sinusoidal wave. The measured motor phase inductance and resistance are 30mH and 10Ω.

Fig. 6.2 shows the experimental setup for the PM spindle motor drive. The insulated gate bipolar transistors (IGBTs) are used as switching devices of the inverter. The switching frequency, in the experiment, is set at $15kHz$. The controller, TMS320F2407 DSP, carries out adaptive full-speed range control algorithms and drives the inverter to generate the applied three-phase voltage. Both of them are given in Fig. 6.3. A dynamometer coupled to the motor as shown in Fig. 6.4 is used as a load, which has the maximum torque 100lb.in(11.3N.m). The measured instrument is illustrated in Fig. 6.5.

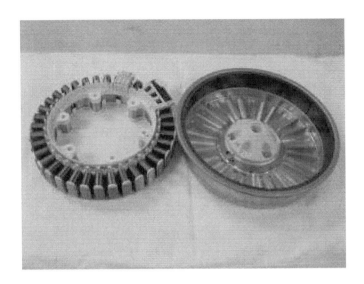

Fig. 6.1 48-pole 36-slot PM spindle motor

Fig. 6.2 Experimental setup of PM spindle drive system

Fig. 6.3 DSP Controller and inverter

Fig. 6.4 Test bench

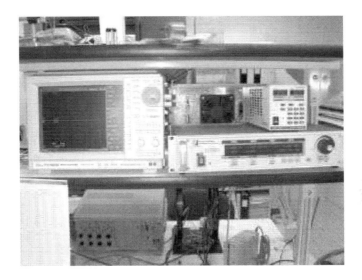

Fig. 6.5 Instruments used for the measurements

6.3 Hardware Setup of the Control System

The hardware structure of the control system is mainly composed of two interconnected modules: TMS320LF2407 EVM board and DMC1500 drive board.

The overall hardware architecture of control system is shown in Fig.6.6a. TMS320F2407 DSP is used as the central processor of the control system to implement full speed range algorithms. The stator currents i_{as}, i_{bs} are sampled and transmitted to DSP thorough DMC1500. The transitions of two Hall Effect signals are detected by capture unit and provide the absolute rotor position every 90 electrical degree. Based on these information, the control system generates the required six PWM gating signals to drive the PM spindle motor.

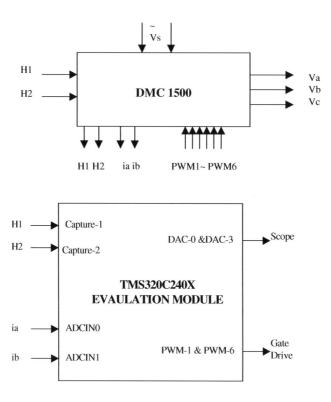

Fig. 6.6a. Overall hardware architecture of the control system.

The Capture units of TMS320f2407 enable logging of the transitions on capture input pins. The Capture units 1,2 associated with EVA is used. Each capture unit is associated with a capture input pin. Each EVA capture unit can choose a GP timer 2 or 1 as its timer base. After a capture unit is enabled, a specified transition on the associated input pin causes the counter value of the selected GP timer to be loaded into the corresponding FIFO stack. At the same time, if there is already one or more valid capture values stored in the FIFO stack the corresponding interrupt flag is set. If the interrupt for that particular vector is unmasked, a peripheral interrupt request is generated.

A user can specify the transition detection of either rising edge, or falling edge or both edges. All capture inputs are provided with Schmitt-triggered input and are synchronized with the CPU clock. In this experiment, two outputs of the Hall effect sensors are interfaced with these capture pins Capture-1 and Capture-2. All the capture pins used are set to detect both rising and falling edges. Hence, every 90 electrical degree of the motor rotation one capture interrupt is generated at EVA by Capture-1 and Capture-2.

TMS320LF2407 has a 10-bit Analog-to-Digital Converter (ADC) with 8 or 16 multiplexed input channel, which transfer the 0~3.3V analog signal to 0-0FFC0h digital signal. For every conversion, any one of the available 16 input channels can be selected through the analog multiplexer. After the conversion is completed, the digital value of the selected channel is stored in the appropriate result register (RESULTn). The minimum conversion time for each of the ADC is 500 ns. The start of conversion can be synchronized with a trigger generated by event managers.

ADCIN0 and ADCIN1 are selected to be input channels of current sensing in the experiment. The trigger to start conversion is underflow interrupt. This function filters the switch frequency when the phase current is read.

TMS320LF240X evaluation board has four DAC channels. The DAC available on the evaluation board are used to display various system variables on oscilloscope in real time. This subroutine accepts the address pointers for four different system variables and then automatically updates the output of DACs to reflect the change in these variables. This feature is very useful during development stage for real time debugging and verification of software.

DMC 1500 is designed to be used as a digital motor controller development system with TMS320F240/243/2407 evaluation board. It allows development to be done with AC induction, Brushless DC, and Switch reluctance motor. Some key features of the DMC1500 are [61]:

- Rated for Bus voltage of 350VDC

- Rated current is 5 amps continuous and 10 amps peak.

- Onboard control power supply (115~230VAC)

- Phase current, Phase voltage, Bus current and voltage sensing

- Hall sensor/Encoder inputs.

The DMC 1500 supports reading the phase current in the lower transistor leg. The currents are measured across three 0.04 ohm resistor. These sensor signal are then filtered for a 40KHz cutoff frequency, clamped to the rails and applied to the non inverting input to an op-amp. For each input, there is also a jumper to apply an offset voltage. The gain of each amplifier can be adjusted from 10 to 111. Thus a bipolar current can be adjusted to the required value of A/D input channels.

Since the rated voltage and current of PM spindle motor is 220V and 2.5A. The above features satisfy all the requirements of PM spindle motor control system.

6.4 Software Development of PM Spindle Motor Control System

DSP-based software is developed to implement the control algorithms of the PM spindle motor drive. During the development, the hardware structures and its specific features are considered in order to achieve high processing speed and precision in the overall control system.

The total control programs of the PM spindle motor includes four modules:

1. Initialization procedure;
2. ADC module;
3. DAC module;
4. Hall Effect signals observation;
5. Full-speed range control module.

Initialization procedure performs memory allocations for the program variables and peripherals, and configures the DSP for specific tasks, etc. ADC is responsible for sampling stator currents, and DAC output the program variable to evaluate the developed program. Hall effect observation module is developed to capture the transition of Hall signal. As the crucial part of the whole program, full-speed range control module achieves the full speed range current control algorithms, and implements speed control and current regulations. The last two modules adopt the interrupt service subroutine.

The flowchart of the overall control system is shown as follows:

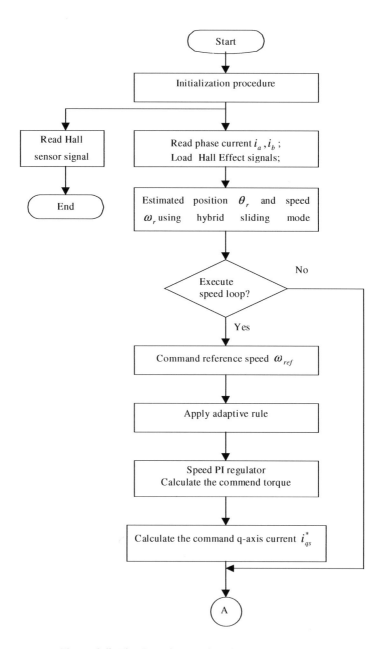

Figure 6.6b The flow chart of the whole control system

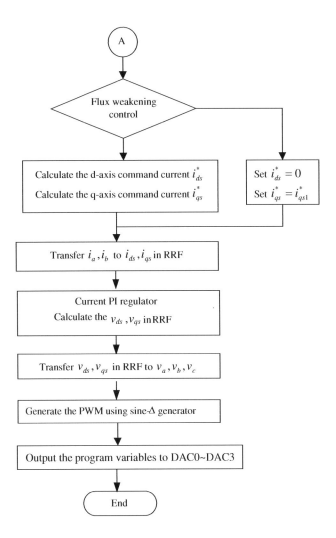

Figure 6.6b Continued

6.5 Experimental Result Analysis

Using experimental setup and programs described above, A large number of experiments were conducted in order to closely examine and verify the basic operation and performance characteristics of the proposed drive. The motor can run using minimum one Hall sensor and also two Hall sensors. For these two cases, the results are obtained under the same conditions except that the numbers of Hall sensors differ. The details of the experiments along with typical waveforms observed are reported in this section.

6.5.1 Hall Signal Identification

Typically PM motor has three Hall sensor signals which their rising edges are aligned with the three-phase or line-to-line back-EMF such that the maximum torque is generated and the torque ripple is reduced. Since there are only two Hall sensor signals in which every two interval is 90 electrical degrees, as shown in Fig. 6.7 for proposed motor, the identification of Hall sensor signals with respect to the corresponding motor phase back-EMF becomes necessary in order to detect absolute rotor position through Hall sensor signal, which is used to correct the output of the sliding mode observer and build the simple observer.

In order to achieve this, the motor is rotated as a generator using another DC motor and the position sensor outputs and phase back-EMF are observed on the oscilloscope. Fig. 6.8 shows line-line back-EMF. It is sinusoidal wave and form three-phase balance system. Thus, the phase back-EMF can be obtained:

$$e_a = \frac{1}{3}(e_{ab} - e_{ca})$$
$$e_b = \frac{1}{3}(e_{bc} - e_{ab}) \qquad (6.1)$$
$$e_c = \frac{1}{3}(e_{ca} - e_{bc})$$

117

Fig. 6.9 and Fig. 6.10 show Hall sensor signals with associated phase back-EMF. It is observed that the Hall signal is aligned with A-phase back-EMF. Some position error is also observed between phase A and Hall sensor signal. It is the arrangement error and should be taken into consideration during design process.

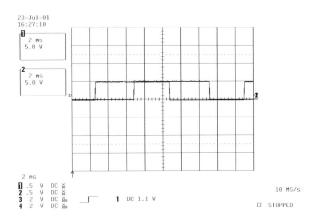

Fig. 6.7 The Hall sensor signals

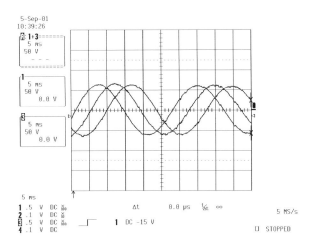

Fig. 6.8 The line-to-line back-EMF

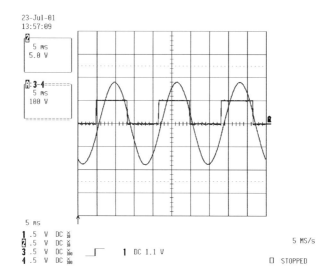

Fig. 6.9 Hall sensor I signal with phase A back-EMF

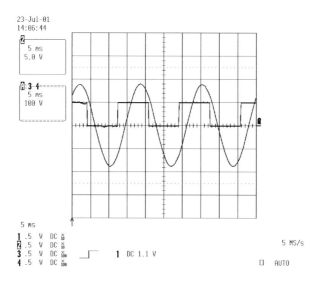

Fig. 6.10 Hall-sensor II signal with phase C back-EMF

6.5.2 Starting Process

Starting is always a problem for surface mount PMSM if the sensorless technology is used. None of the methods can start the motor with full load. By employing the Hall sensor signal, the rotor position can be available from the beginning such that the vector control algorithm can be implemented even for starting.

Fig. 6.11 and Fig. 6.12 show the four actual motor positions that come from Hall sensors signal in one cycle and estimated rotor position by simple observer. It can be seen that the simple observer performs well at 5 rpm and position error is acceptable at 1 rpm. The motor cannot run smoothly at very low speeds without load due to the cogging torque. If the load condition is considered, it can filter the cogging torque.

Starting process with 11.3 N.m load (90% full load and maximum output from dynamometer) is shown in Fig. 6.13 and Fig. 6.14 with one-Hall sensor and two-Hall sensor, respectively. It is clear that vector control is implemented from the beginning as illustrated in d and q-axes currents. Thus, for both of the cases the system has a faster speed response and much less current overshoot compared to the conventional open loop starting in sensorless methods. The starting current is less than 30% of the rated current, which results in less current limit protection in power electronics side so that it is possible to reduce the inverter size. Comparing both cases, the case with two-Hall sensor has better performance. It has almost twice faster speed response as that of the case with one-Hall sensor, less overshoot so that the phase current and d and q-axis currents in rotor reference frame have been much less disturbed. If the number of Hall sensors increases, a much better performance is expected.

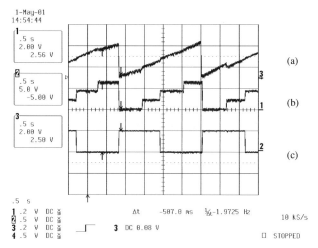

(a) estimated position   (b) Four real position from Hall sensors
(c) One Hall sensor signal

Fig. 6.11 Estimation position and Hall signals at 1 rpm

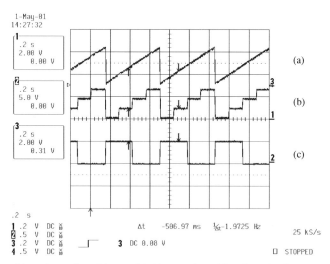

(b) estimated position   (b) Four real position from Hall sensors
(c) One Hall sensor signal

Fig. 6.12 Estimation position and Hall signals at 5 rpm

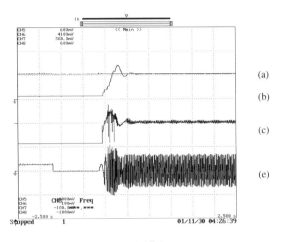

time(0.5s/div)

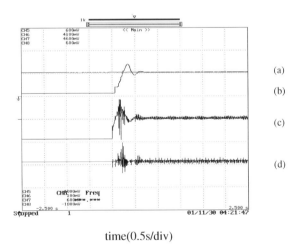

time(0.5s/div)

(a) command speed (b) actual speed
(c) q-axis current in rotor reference frame (d) d-axis current in rotor reference frame
(e) A-phase current

Fig. 6.13 Starting process using one Hall sensor with 11.3N.m load

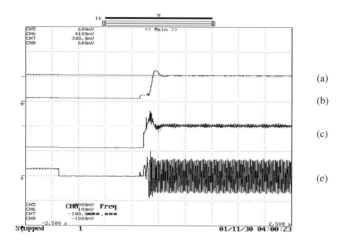

time(0.5s/div)

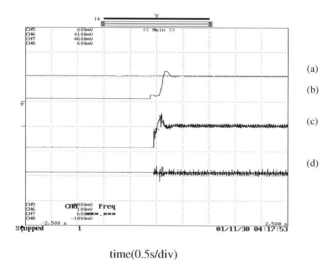

time(0.5s/div)

(a) command speed (b) actual speed
(c) q-axis current in rotor reference frame (d) d-axis current in rotor reference frame
(e) A-phase current

Fig. 6.14 Starting process using two Hall sensors with 11.3N.m load

6.5.3 Hybrid Sliding Mode Observer

Hybrid sliding mode observer is used to estimate rotor position. It consists of three parts, which are sliding mode current observer, position estimator and Hall correction. Based on the analysis of the Chapter III, the output of sliding mode current observer is the sliding function and can be considered as the back-EMF with high frequency harmonics. By filtering this harmonics, the smooth back-EMF is obtained, which provide the position information. Since the numbers of Hall sensors do not affect much the sliding mode observer's performance, for case with one-Hall sensor and case with two-Hall sensor, the performance is almost the same. Thus, the following figures are based on two Hall sensor experiments.

From the above structure of sliding mode observer, two parameters, sliding gain and cut-off frequency, mainly affect the performances of the observer. Their effects are investigated and shown in Fig. 6.15. Since the sliding gain should be larger than back-EMF. For these figures, the DC-bus voltage is assumed as the base value of sliding gain and the motor runs at 10 rpm. Table 6.1 gives the experimental conditions.

Table 6.1 Experimental conditions

	Fig. 6.15(a)	Fig. 6.15(b)	Fig. 6.15(c)	Fig. 6.15(d)
Cut-off frequency	5 Hz	50Hz	50Hz	5Hz
Sliding gain	1/8	1/8	3/4	3/4

It can be observed that the observer has the better performance while the appropriate sliding gain and cut-off frequency is selected as shown in Fig. 6.15 (a). Since the sliding function behaves like a varying-frequency PWM generator in which the high frequency harmonics is close the fundament wave, the cut-off frequency close to the fundamental frequency results in the good performance even though the sliding gain is large as shown in Fig. 6.15(d). If the cut-off frequency is far from the fundamental frequency and proper sliding gain is selected, the waveform becomes worse as shown in Fig. 6.15(b). The worst waveform as shown in Fig.6.15 (c) is obtained if the cut-off frequency and sliding gain is selected incorrectly.

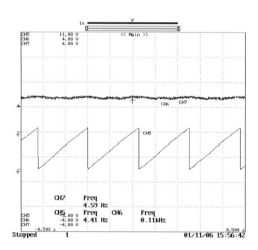

From top to bottom: motor current (stationary frame)
                       Estimated current from current observer
                       Estimated position

(a)

Fig. 6.15 The effects of parameters on sliding mode observer

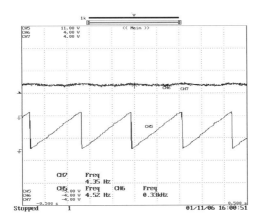

From top to bottom: motor current (stationary frame)
                                Estimated current from current observer
                                Estimated position

(b)

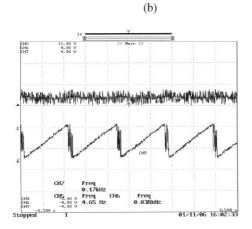

From top to bottom: motor current (stationary frame)
                                Estimated current from current observer
                                Estimated position

(c)

Fig. 6.15 Continued

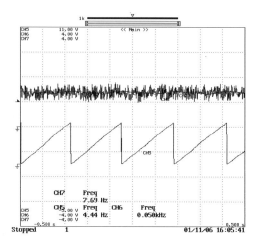

From top to bottom: motor current (stationary frame)
                      Estimated current from current observer
                      Estimated position

(d)

Fig. 6.15 Continued

The above analysis provides an idea on how to design the sliding mode observer. However, since the filtered back-EMF has the phase delay with the real back-EMF, the compensation algorithm should be developed for position estimator. If the cut-off frequency varies, it is difficult to design compensation algorithm. Another problem is the accumulating error for observer based methods. Hall sensor correction part gives a solution for these problems. By correcting the estimated position every cycle, no compensation algorithm is required and accumulating error is also eliminated. Fig. 6.16 evaluated sliding mode current observer. It is clear that the estimated current track the motor real current, which means the sliding function is back-EMF with high frequency harmonics and ensure the correction of position estimation.

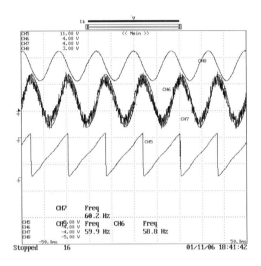

From top to bottom: estimated back-EMF, real motor current
estimated motor current, estimated rotor position

Fig. 6.16 Performance of sliding mode current observer

Following the above concept, hybrid sliding mode observer is evaluated from medium speed to high speed. Fig. 6.17 - 6. 20 display the estimated position in different motor speed. All these figures prove the performance of the hybrid sliding mode observer.

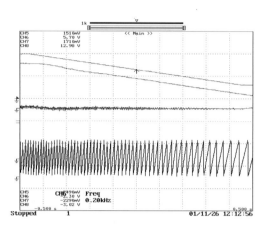

From top to bottom: commanded speed, real rotor speed
q-axis current, estimated rotor position

Fig. 6.17 Performance of the hybrid sliding mode observer

when speed varies from 30Hz to 90Hz

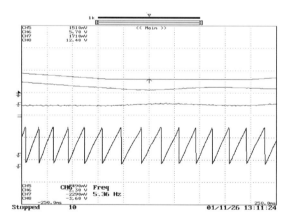

From top to bottom: commanded speed, real rotor speed
q-axis current, estimated rotor position

Fig. 6.18 Performance of the hybrid sliding mode observer at 30Hz

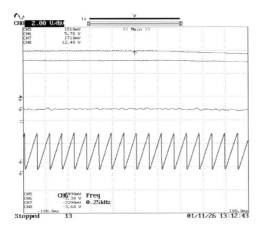

From top to bottom: commanded speed, real rotor speed
q-axis current, estimated rotor position

Fig. 6.19 Performance of the hybrid sliding mode observer at 90Hz

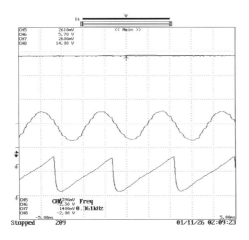

From top to bottom: commanded speed, real rotor speed
A-phase current, estimated rotor position

Fig. 6.20 Performance of the hybrid sliding mode observer at 360Hz

6.5.4 Maximum Torque/Current Ratio Operation

As described in Chapter IV, the constant torque operation developed for PM spindle motor involves both speed and current controls. The speed regulation is used to control torque and correspondingly provides the torque components of the stator currents, which is q-axis current in rotor reference frame (RRF). The d-axis current in RRF is set to zero to achieve the maximum torque/current ratio. The current loops make the stator actual d and q-axis currents in RRF track the commanded d and q-axis current obtained in the above description.

In this experiment, several characteristics of constant torque operation are tested with the PM spindle motor: 1) system performance under speed variation; 2) system performance under load variation; 3) accelerating and decelerating performance; These features are evaluated for case of one-Hall sensor and the case of two-Hall sensor.

The dynamic response of speed, q-axis current in RRF and phase current with 11.3N.m load is shown in Fig.6. 21 and Fig. 6.22. Two Hall sensors are used in this case. Clearly, the actual speed is capable of following the reference speed very well in the tested range from 20Hz to 80Hz. With the gradual increase of speed, smooth change of stator current and q-axis current in RRF are observed during the dynamic state. At the steady state, the amplitude of stator current is essentially kept unchanged

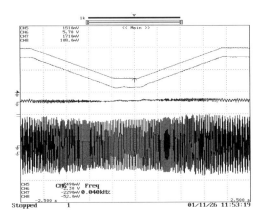

From top to bottom: Commanded Speed   Actual Speed
q-axis current in RRF   A-phase current

Fig. 6.21 Motor dynamic response when the speed changes smoothly with full load

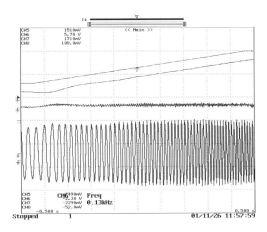

From top to bottom: Commanded Speed   Actual Speed
Q-axis current in RRF   A-phase current

Fig. 6.22 Highlight of the motor dynamic response when speed change smoothly with full load

Fig. 6.23 and 6.24 show that speed suddenly changes from 20Hz to 80Hz and from 80Hz to 20 Hz with 6 N.m load respectively. It can be observed that the q-axis current immediately varies a lot to force the real speed follow the commanded speed, and d-axis current is almost kept at zero with the disturbance because d-axis and q-axis current are controlled in dependently in vector control algorithm. After very short time, motor speed tracks the commanded speed and q-axis current remains constant again with little decreasing for speed-down and increasing for speed-up.

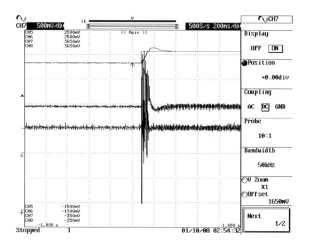

From top to bottom: commanded speed   motor real speed
                       q-axis current in RRF   d-axis current in RRF

Fig. 6.23 The dynamic response when the speed suddenly varies from 20Hz to 80Hz

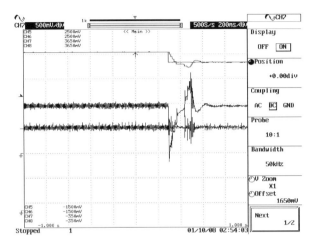

From top to bottom: commanded speed   motor real speed
                      q-axis current in RRF d-axis current in RRF

Fig. 6.24 The dynamic response when the speed suddenly varies from 80Hz to 20Hz

The load rejection is shown in Fig. 6.25 (a),(b). When the 6 N.m load is suddenly added to the motor as shown in Fig. 6.25 (a), the motor speed drops less than 5 rpm and then tracks the commanded value. At the same time, q-axis current in RRF frame increases to reflect the load variation. The d-axis remains zero such that the maximum torque/current ratio is achieved even in dynamics response. The q-axis current wave outlines the wave of phase-A current, which shows that the d and q current is decoupled for vector control. On the other hand, the control system performs similar when a 6N.m load is suddenly removed from the motor.

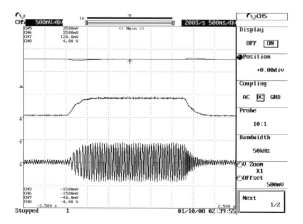

From top to bottom: Commanded speed   motor actual speed

q-axis current,   phase-A current

(a)

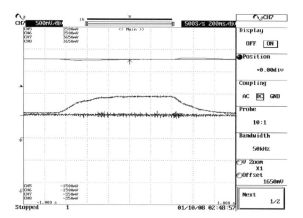

From top to bottom: commanded speed   motor actual speed

q-axis current in RRF,   d-axis current in RRF

(b)

Fig.6.25 The load rejection when 6 N.m load is suddenly added to and removed from the motor

Fig. 6.26 and 6.27 display the load rejection for case with one-Hall sensor and case with two-Hall sensors, respectively when speed varies from 20Hz to 80Hz. When 11.3 N.m load (full load) is suddenly added to the motor, both cases perform well for the load rejection as described in last paragraph even the speed is varying. For case with two-Hall sensors, the speed varies two times faster than case with one-Hall sensor.

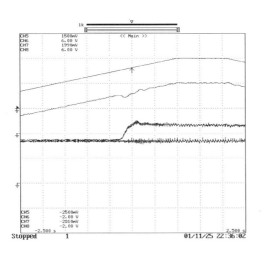

From top to bottom: commanded speed , motor actual speed
q-axis current in RRF, d-axis current in RRF

Fig. 6.26 The load rejection when 11.3 N.m load is suddenly added to the motor with one-Hall sensor

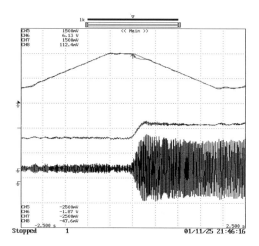

From top to bottom: Commanded speed, motor actual speed
q-axis current, phase-A current

Fig. 6.27 The load rejection when 11.3 N.m load is suddenly added to the motor with two-Hall sensor

Accelerating and decelerating feature is important for washing machine and track application. It is the key index to evaluate system design and controller design. Since vector control is implemented in the total system, a faster accelerating and decelerating feature is expected. Fig. 6.28 and 6.29 illustrate the acceptable feature for case of one-Hall sensor and case with two-Hall sensors, respectively. It can be observed that the motor speed changes from 20 Hz to 80 Hz or from 80Hz to 20 Hz within 1sec under full load condition. It is faster enough for washer application. For case with two-Hall sensors, the motor actual speed varies more smoothly and tracks more accurately than case with one-Hall sensor. If the speed varies twice faster than the above case, as shown in Fig.6.30 and 6.31 for case with one-Hall sensor and case with two-Hall sensors, respectively, the speed tracking become worse and cannot remain constant for constant commanded speed.

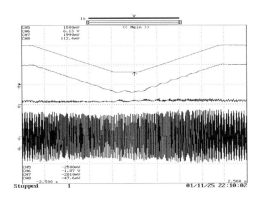

From top to bottom: Commanded speed   motor real speed
                    q-axis current,   phase-A current

Fig. 6.28 Accelerating and decelerating feature under 11.3 N.m load
with one-Hall sensor

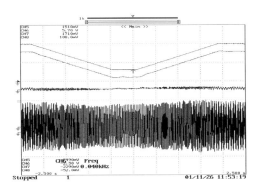

From top to bottom: Commanded speed   motor real speed
                    q-axis current,   phase-A current

Fig. 6.29 Accelerating and decelerating feature under 11.3 N.m load
with two-Hall sensors

138

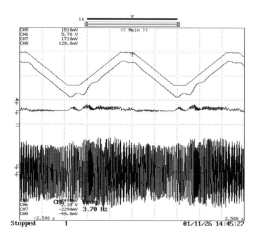

From top to bottom: Commanded speed   motor real speed
                   q-axis current,   phase-A current

Fig. 6.30 Accelerating and decelerating feature in worse case under 11.3 N.m load with one-Hall sensor

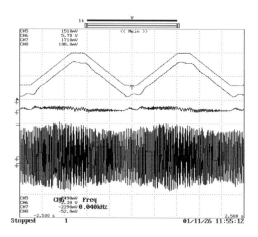

From top to bottom: Commanded speed   motor real speed
                   q-axis current,   A-phase current

Fig. 6.31 Accelerating and decelerating feature in worse case under 11.3 N.m load with one-Hall sensor

6.5.5 Flux Weakening Control

Flux-weakening control is developed to extend the motor speed. For PMAC motor, The amplitude of the back-EMF increases linearly with the rotor speed. Current control (thus torque control) gradually degrades when the speed increases into the regime where the back-EMF approach the amplitude of the DC link source voltage. Eventually the current regulators saturate, losing the ability to force the commanded current into the motor phase.

Fig. 6.32 illustrates this process. In these experiments, the DC bus voltage stays the same. The commanded d-axis current is set to zero. For each case, first the motor run in no load and then the load is added to the motor until q-axis current regulator saturates. It can be seen that the rated current is obtained in 90Hz as shown in (a) of Fig. 6.32, which hits the threshold for constant torque operation, then output current reduces shown in (b) and becomes the minimum current to maintain the motor speed in (c). At this point, the maximum speed is reached but motor can not be loaded.

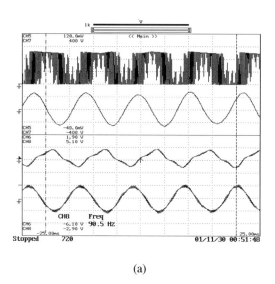

(a)

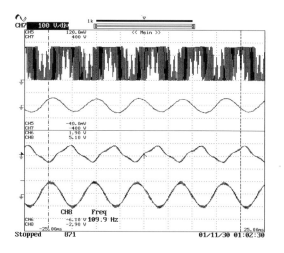

(b)

From top to bottom: phase voltage   phase current
                      Back-iron induced voltage   tooth induced voltage

Fig. 6.32   The system response without flux weakening control
             while motor speed increases

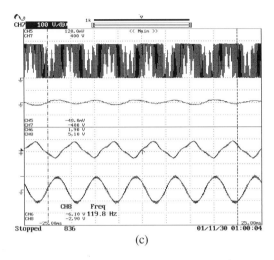

(c)

From top to bottom: phase voltage   phase current
                     Back-iron induced voltage   tooth induced voltage

Fig. 6.32   Continued

Flux weakening algorithm is developed to extended speed range as described in Chapter IV. By adding the negative d-axis stator current (thus negative stator flux in d-axis), the positive d-axis magnet flux is reduced. Thus high-speed operation becomes possible even though the back-EMF is much higher than phase voltage.

Fig. 6.33 illustrates the flux weakening control with voltage and current limitation, which motor speed increases from 90Hz to 360Hz. For all these experiments, the phase voltage is the same and d-axis current is given according to the analysis in 4.32. For each case, first the motor run at no load at given speed and then the load is added to the motor with the same speed until q-axis current regulator saturates. It can be observed that the phase voltage and current limit is reached and the motor speed is extended to 360Hz. The voltage and current vectors are moved along the limit circle which means region II with the optimal current control is achieved. Considering the parameter variation, adaptive speed and current PI are used for the overall system.

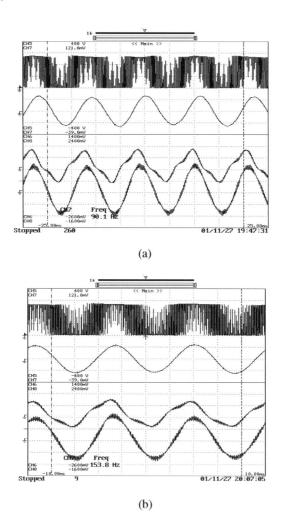

(a)

(b)

From top to bottom: phase voltage   Phase current
                     Back iron induced voltage   Teeth induced voltage

Fig. 6.33 The process of flux weakening control

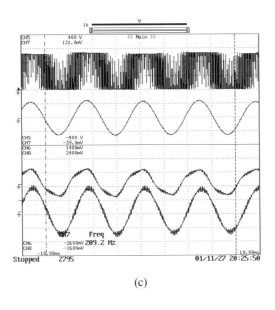

(c)

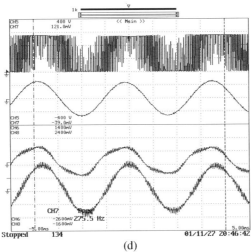

(d)

From top to bottom: phase voltage   Phase current
                     Back iron induced voltage   Teeth induced voltage

Fig. 6.33 Continued

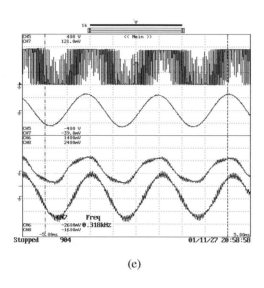

(e)

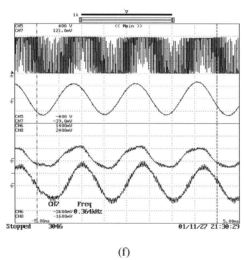

(f)

From top to bottom: phase voltage   Phase current
                     Back iron induced voltage Teeth induced voltage

Fig. 6.33 Continued

Fig. 6.34 - Fig. 6.36 provide the overview of flux weakening control. Speed response is shown in Fig. 6.34. Clearly, the motor actual speed follows the commanded speed. At the same time, negative d-axis current is added to weaken the magnet flux with respect to different speed and the q-axis current stays the same due to no-load operation. Fig. 6.35 displays the speed, torque and power relations in flux weakening control. As the speed increases, the torque decreases and the power stays constant in certain region. As introduced in Chapter IV, since the stator flux is less than magnet flux in d-axis, the power decreases when the speed over the threshold of constant power operation, which is shown in the Fig. 6.35. The relation of phase induced voltage and phase voltage is shown in Fig. 6.36. It can be seen that the induced voltage is not over the one in the threshold point of constant torque operation and almost keep the same value.

One-Hall case is also evaluated in the flux weakening control. For same performance as two-Hall sensor case, the case with two–Hall sensor has almost twice faster speed response than case with one Hall as shown in Fig. 6.37.

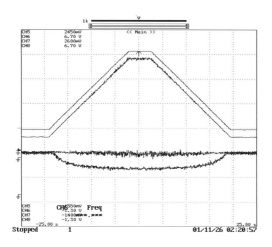

From top to bottom: commanded speed   motor actual speed
                 q-axis current   d-axis current

Fig. 6.34 Speed response of flux weakening control with two-Hall sensor under no load

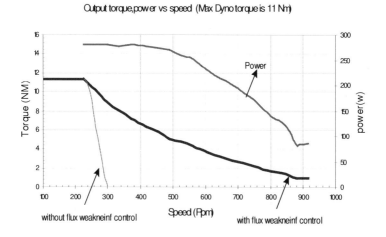

Fig. 6.35 Torque, power vs speed

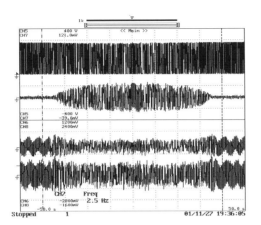

From top to bottom: phase voltage   Phase current
                      Back iron induced voltage   Teeth induced voltage

Fig. 6.36 The relation of phase voltage and phase induced voltage in flux weakening control

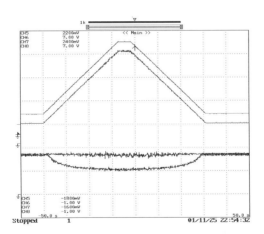

From top to bottom: commanded speed   motor actual speed
                     q-axis current   d-axis current

Fig. 6.37 Speed response of flux weakening control with one-Hall sensor under no load

6.5.6 Washing Machine Application

After evaluating the developed algorithm with dynamometer as the load, the whole technology is implemented in real direct-drive washing machine. Fig. 6.38 shows the prototype direct-drive washing machine. The 16 lb cloth as shown in Fig. 6.39 is added as the load. Fig. 6.40 displays the washer running in the 900rpm with 16 lb load. The drive system of washing machine performs the same as the system is tested with dynamometer. The speed response is given in Fig. 6.41. The similar waveform is observed as the one obtained from dynamometer test, in which the speed tracks the commanded speed, and both d and q-current are decoupled.

Off-balance test is very important in washing machine because the cloth may not be put symmetrically. Fig. 6.42 shows the case that the washer is running in washing cycle after spin cycle. Since the cloths are tighten together and no water is in its basket, it seems off-balance load is given. Clearly, the speed tracking is still very well even with off-balance load.

Fig. 6.38 Direct –drive washing machine

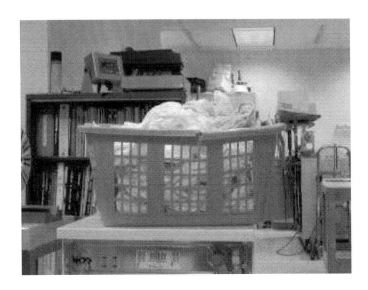

Fig. 6.39 The cloth used in the test

Fig. 6.40 The high speed operation (900rpm) of D-D washer

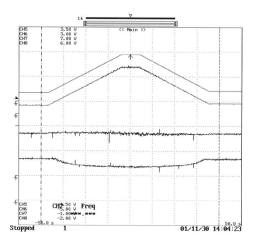

From top to bottom: commanded speed   motor real speed

q-axis current   d-axis current

Fig. 6.41 Speed response of direct drive washer

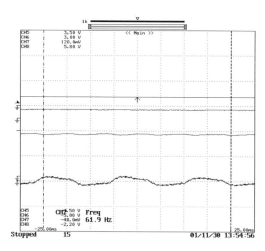

From top to bottom: commanded speed   motor real speed

q-axis current   A-phase current

Fig. 6.42 Off-balance operation of direct drive washer

CHAPTER VII

CONCLUSIONS AND FURTHER WORK

In this chapter, a summary of the work accomplished in the modeling and development of the full-speed range PM AC motor systems is presented. Suggestions for future research in this area are also proposed.

7.1 Conclusions

This dissertation presents full speed range control scheme for PMAC motors using hybrid sliding mode observer. The proposed method for sinusoidal PM motor achieves full load starting, maximum torque/current ratio below rated speed, and extended motor speed up to six times of the rated speed. On the other hand, the developed method for trapezoidal PM motor results in an increased developed torque and also shows the same controllability as in the sinusoidal PM motor over the flux weakening operation comparing to the conventional six-step current control.

Using finite element analysis, the three main characteristics of 48-pole 36-slot have been investigated. It is found that the sinusoidal back-EMF can be generated in PM spindle motor by the increase of the number of pole and proper design of the slot opening. After that, since the stator poles and rotor poles are different for this sinusoidal back-EMF spindle motor, torque generation with the different stator poles and rotor poles is first studied. Unlike the conventional three-phase current, it is found that the different motor frequency and phase shift between two phases should be used to generate positive torque. Finally, the flux-weakening capability of the proposed motor is evaluated. The flux distribution wave shows the magnetic material is still in recovery region after motor speed is extended to six times. These analyses provide the basis for the further study.

Hybrid sliding mode observers are studied for both sinusoidal back-EMF and trapezoidal back-EMF PM motors to provide the rotor position. By employing at least one low-cost Hall sensor, a simple observer is developed to provide rotor speed during starting and low speed operation. From the medium to high speed, the sliding mode observer is used to estimate the position. Hall sensor signal is also used to correct the estimated position obtained from sliding mode observer and improve its performance. The proposed methods have all the advantages of sliding mode observer and eliminate its drawback.

The d-q model in rotor reference frame is proposed for the PM spindle motor using new transformation based on the motor analysis. A full speed range control algorithm is developed using this d-q model. Adaptive technique is applied over the full speed range considering the parameter variation. The developed scheme has used vector control from zero speed. It achieves full load starting with 0.5 s dynamic response and less than 30% current overshot (worst 3 times current overshot for open loop starting), maximum/current ratio for constant torque operation, and extend motor speed up to six times of rated speed with optimal current control in flux weakening region.

A full-speed range control algorithm of trapezoidal back-EMF PM motor is also studied. The developed topology looks like field-oriented control for sinusoidal PM motors but the fifth and seventh harmonic variables have been taken into consideration as well as the fundamental variables. At every instant of time all of the three phases conduct, and every switch conducts for 180 degree. Thus, it increases the motor torque during constant torque operation and exhibits the same controllability as the sinewave PM motor during flux weakening operation. In other words, by controlling the d and q-axes currents in RRF, a given point with certain speed and torque in speed vs torque characteristics can be reached.

The full-speed range control schemes developed for sinusoidal back-EMF and trapezoidal back-EMF PM motor have been simulated using MATLAB/SIMULINK. A prototype of the PM spindle motor and the control system based on TMS320C2407 DSP have been built and experimented in the Whirlpool Corporation. The PM spindle motor system developed achieves good dynamic and steady-state performance for overall speed range even the full load is suddenly added to the system and speed varied between 20Hz and 80Hz with half load. The experimental results obtained validate the theoretical analysis and simulation results.

Finally the developed algorithm is implemented on a direct drive washing machine. The load test and off-balance test have been done with 16 lb cloth. The desired performance has been obtained.

7.2 Suggestions for Further Research

For further research in this area, the following suggestions are made

- It has been proved that the proposed PM spindle motor has very good flux-weakening performance even with surface mount rotor. Generally interior PM motor demonstrates better flux-weakening capability than surface mount PM motor. It is desirable to investigate which structure of PM motor has the best flux-weakening capability and furthermore develop optimal design of PM motor for flux weakening operation.

- The implementation of full-speed range control for BLDC motor with good flux weakening capability and then the performance is compared with the conventional six-step current control.

- For the purpose of application, the brake system can be added to the whole control system.

REFERENCES

[1] T.J.E. Miller, *Brushless PM and Reluctance Motor Drives*, Oxford, UK: Clarendon, 1989.

[2] J.R. Hendershot, and T.J.E. Miller, Design *of Brushless Permanent Motor*, Oxford, UK: Clarendon, 1994.

[3] B.K.Bose, *Power Electronics and Variable Frequency Drives*, Piscataway, NJ: IEEE Press, 1996.

[4] P. Pillay and R. Krishnan, "Modeling, simulation, and analysis of permanent-magnet drives, part I: the permanent-magnet synchronous motor drive," *IEEE Trans. Ind. Applicat.*, vol. 25, pp. 265 –273, March/April, 1989.

[5] P. Pillay and R. Krishnan, "Modeling, simulation, and analysis of permanent-magnet drives, part II: the brushless DC motor drive," *IEEE Trans. Ind. Applicat.*, vol. 25, pp. 274 –279, March/April, 1989.

[6] N.Boules; "Prediction of no-load flux distribution in permanent magnet machines", *IEEE Trans. Ind. Applicat.*, vol. 21, pp. 633-643, 1985.

[7] T, Sebastian and V, Gangla. "Analysis of induced EMF and torque waveforms in a brushless permanent magnet machine," in *Conf. Rec. IEEE-IAS Annu. Meeting*, vol. 1, 1994, pp. 240 -246.

[8] T.J.E. Miller and R. Rabinovici, "Back-EMF waveforms and core losses in brushless DC motors", *IEE Proc. Electr. Power Appl.*, vol. 141, 1994, pp. 144-154.

[9] C. Studer, A, Keyhani, T. Sebastian, and S. K. Murthy, "Study of cogging torque in permanent magnet machines," in *Conf. Rec. IEEE-IAS Annu. Meeting*, vol. 1, 1997, pp. 42 –49.

[10] Z. J. Liu; C. Bi; H. C. Tan, and T.S. Low, "Modelling and torque analysis of permanent magnet spindle motor for disk drive systems," *IEEE Trans. Magn.*, vol. 30, pp. 4317-4319, November, 1994.

[11] S.M. Hwang, K. T. Kim, W. B. Jeong, Y.H. Jung, and B. S. Kang, "Comparison of vibration sources between symmetric and asymmetric HDD spindle motors with rotor eccentricity," *IEEE Trans. Ind. Applicat.*, vol. 37, pp. 1727 –1731, November/December, 2001.

[12] Y.D. Yao, D.R. Huang, J.C. Wang, S.J. Wang, T.F. Ying, and D.Y. Chiang, "Study of a high efficiency and low cogging torque spindle motor," *IEEE Trans. Magn.*, vol. 34, pp. 465 –467 Mar., 1998.

[13] S.X. Chen, T.S. Low, H. Lin, and Z.J. Liu, "Design trends of spindle motors for high performance hard disk drives," *IEEE Trans. Magn.*, vol. 32, pp. 3848 –3850, Sept. 1996.

[14] A.B. Kulkarni; and M. Ehsani, "A novel position sensor elimination technique for the interior permanent-magnet synchronous motor drive," *IEEE Trans. Ind. Applicat.*, vol. 28, pp. 144 –150 , January/February, 1992 .

[15] M.J. Corley, and R.D. Lorenz, "Rotor position and velocity estimation for a salient-pole permanent magnet synchronous machine at standstill and high speeds," *IEEE Trans. Ind. Applicat.*, vol. 34, pp. 784 -789, July/August, 1998.

[16] J. M. Kim, S. J. Kang, and S. K. Sul, "Vector control of interior permanent magnet synchronous motor without a shaft sensor," in *Proc. IEEE APEC'97*, vol. 2, 1997, pp. 743 -748.

[17] J.P. Johnson, M. Ehsani, and Y. Guzelgunler, "Review of sensorless methods for brushless DC," *in Conf. Rec. IEEE-IAS Annu. Meeting*, vol. 1, 1999, pp. 143 - 150.

[18] R. Wu, and G. R. Slemon, "A permanent magnet motor drive without a shaft sensor," *IEEE Trans. Ind. Applicat.*, vol. 27, pp. 1005-1011, September./October, 1991

[19] M. Naidu and B.K. Bose, "Rotor position estimation scheme of a permanent magnet synchronous machine for high performance variable speed drive," in *Conf. Rec. IEEE-IAS Annu. Meeting*, vol. 1, 1992, pp. 48 – 53.

[20] H. Watanabe, H. Katsushima, and T. Fujii, "An improved measuring system of rotor position angles of the sensorless direct drive servomotor," in *Proc. IEEE IECON'91*, vol. 1, 1991, pp. 165–170.

[21] J. S. Kim and S. K. Sul, "New approach for high-performance PMSM drives without rotational position sensors," *IEEE Trans. Power Electron.*, vol. 12, pp.904–911, Sept. 1997.

[22] R. Dhaouadi, N. Mohan, and I. Norum, "Design and implementation of an extended Kalman filter for the state estimation of a permanent magnet synchronous motor," *IEEE Trans. Power Electron.* ,vol. 63, pp. 491-497., July 1991.

[23] A. Bado, S. Bolognani and M. Zigliotto "Effective estimation of speed and rotor position of a PM synchronous motor drive by a Kalman filtering technique," in *Proc. IEEE PESC'92*, vol. 2, 1992, pp. 951–957.

[24] T. Low, T. Lee, and K.Chang, "A nonlinear speed observer for permanent-magnet synchronous motor," *IEEE Trans. Ind. Electron.*, vol. 40, pp. 307-315,June 1993.

[25] R.B. Sepe and J. H. Lang, "Real-time observer-based (adaptive) control of a permanent-magnet synchronous motor without mechanical sensors," *IEEE Trans. Ind. Applicat.*, vol. 28, pp. 1345-1352, November/December. 1992

[26] J. S. Kim and S. K. Sul, "High performance PMSM drives without rotational position sensors using reduced order observer," in *Conf. Rec. IEEE-IAS Annu. Meeting*, vol. 1, 1995, pp. 75-82.

[27] T. Senjyu, M. Tomita, S. Doki, and S. Okuma, "Sensorless vector control of brushless DC motors using disturbance observer," in *Proc. IEEE PESC'95*, vol. 1, 1995, pp. 772 - 777.

[28] J. Hu, D. Zhu; Y. Li, and J. Gao, "Application of sliding observer to sensorless permanent magnet synchronous motor drive system," in *Proc. IEEE PESC'94*, vol.1, 1994, pp. 532 - 536.

[29] Y. S, Han, and Y. S. Kim, "The speed and position sensorless control of PMSM using the sliding mode observer with the estimator of stator resistance," in *Proc. IEEE IENCON 99,* vol. 2, 1999, pp. 1479 –1482.

[30] A. Consoli, G. Scarcella, and A. Testa, "Sensorless control of PM synchronous motors at zero speed," in *Conf. Rec. IEEE-IAS Annu. Meeting*, vol. 2, 1999, pp. 1033 - 1040.

[31] J. I. Ha, K. Ide, T. Sawa, and S. K. Sul," Sensorless position control and initial position estimation of an interior permanent magnet motor," in *Conf. Rec. IEEE-IAS Annu. Meeting*, vol. 4, 2001, pp. 2607 -2613

[32] D. W. Chung, J. K. Kang, and S. K. Sul, "Initial rotor position detection of PMSM at standstill without rotational transducer," in *Conf. IEMD '99*, 1999, pp. 785 –787

[33] T.A Nondahl, G. Ray, and P. B. Schmidt, "A permanent magnet rotor containing an electrical winding to improve detection of motor angular position," in *Conf. Rec. IEEE-IAS Annu. Meeting*, vol. 1, 1998, pp. 359 – 363.

[34] H.A.Toliyat, L. Hao, D.S. Shet, and T.A. Nondahl, "Position-sensorless control of surface-mount permanent-magnet AC (PMAC) motors at low speeds," *IEEE Trans. Ind. Electron.*, vol. 49 , pp. 157 – 164, Feb. 2002.

[35] J. Bu, L. Xu, T. Sebastian, and B.Liu, "Near-zero speed performance enhancement of PM synchronous machines assisted by low-cost Hall effect sensors," in *Proc. IEEE APEC'98*, vol. 1, 1998, pp. 64 - 68.

[36] K. A. Corzine and S.D. Sudhoff, "A hybrid observer for high performance brushless DC motor drives ," *IEEE Trans. Energy Conversion*, vol. 11 , pp. 318 – 323, June 1996

[37] B.Sneyers, D.W. Novotny, and T.A. Lipo, " Field-weakening in buried permanent magnet AC motor drives," *IEEE Trans. Ind. Applicat.*, vol.21, pp. 398-407, March/April 1985 .

[38] T. M. Jahns, " Flux-weakening regime operation of an interior permanent magnet motor drive," *IEEE Trans. Ind. Applicat.*, vol.23, pp. 681-689,July/August 1987.

[39] R.F. Schiferl and T.A. Lipo, "Power capability of salient pole permanent magnet synchronous motors in variable speed drive applications," *IEEE Trans. Ind. Applicat.*, vol.26, pp. 115-123, January.-February. 1990.

[40] S. Morimoto, Y. Takeda, T. Hirasa, and K. Taniguchi, "Expansion of operating limits for permanent magnet motor by optimum flux-weakening," in *Conf. Rec. IEEE-IAS Annu. Meeting*, vol. 1, 1989, pp. 51 – 56.

[41] A. K. Adnanes and T.M.Undeland, "Optimum torque performance in PMSM drives above rated speed," in *Conf. Rec. IEEE-IAS Annu. Meeting*, vol. 1, 1991, pp. 169 – 175.

[42] A.K. Adnanes, "Torque analysis of permanent magnet synchronous motors," in *Proc. IEEE PESC'91*, vol.1, 1991, pp. 695 - 701.

[43] W. L. Soong and T.J.E. Miller," Theoretical limitations to the field-weakening performance of the five classes of brushless synchronous AC motor drive," in *Conf. IEMD '93*, 1993, pp. 127 –132.

[44] W. L. Soong and T.J.E. Miller, "Practical field-weakening performance of the five classes of brushless synchronous AC motor drive," in *Proc. IEEE PESC'93*, vol.5, 1993, pp. 303 - 310.

[45] R. Krishnan, "Control and operation of PM synchronous motor drives in the field-weakening region", in *Proc. IEEE IENCON 93.* vol. 2 , 1993 , pp. 745 -750

[46] N. Bianchi, S. Bolognani, and M. Zigliotto, "Analysis of PM synchronous motor drive failures during flux weakening operation," in *Proc. IEEE PESC'96*, vol.2, 1996, pp. 1542 - 1548.

[47] T.M. Jahns, "Torque production in permanent-magnet synchronous motor drives with rectangular current excitation," *IEEE Trans. Ind. Applicat.*, vol.20, pp. 803-813.July/August 1984.

[48] R.C. Becerra and M. Ehsani, "High-speed torque control of brushless permanent magnet motors," *IEEE Trans. Ind. Electron.*, vol. 35 , pp. 402 –406, Aug. 1988.

[49] S.K. Safi, P.P. Acarnley, and A.G. Jack,"Analysis and simulation of the high-speed torque performance of brushless DC motor drives ," *in Proc IEE EPA'95* 2 3 , May 1995 , vol. 142, pp. 191 -200

[50] S. P. Hong, H. S. Cho, H. S. Lee, H. R. Cho, and H.Y. Lee, "Effect of the magnetization direction in permanent magnet on motor characteristics," *IEEE Trans. Magn.*, vol. 35 , pp. 1231-1234, May 1999.

[51] J. E. Slotine, J. K. Hedrik and E.A. Miasawa, " On sliding observers for nonlinear systems," *ASME J.Dyn. Syst. And Meas*, Vol.109 ,Sept. ,1987, pp. 245-252,

[52] V.I. Utkin, *Sliding in Control and Optimization*, Berlin: Springer-Verlag, 1992.

[53] I. Husain, S. Sodhi, and M. Ehsani, "A sliding mode observer based controller for switched reluctance motor drives," in *Conf. Rec. IEEE-IAS Annu. Meeting*, vol. 1, 1994, pp. 635-643.

[54] R.A. McCann and I. Husain, "Application of a sliding mode observer for switched reluctance motor drives," in *Conf. Rec. IEEE-IAS Annu. Meeting*, vol. 1, 1997, pp. 525-532.

[55] P.L. Chapman; S.D. Sudhoff, and C.A. Whitcomb, "Multiple reference frame analysis of non-sinusoidal brushless DC drives," *IEEE Trans. Energy Conversion*, vol. 14, pp. 440-446, Sept. 1999.

[56] S.D. Sudhoff, "Multiple reference frame analysis of an unsymmetrical induction machine," *IEEE Trans. Energy Conversion*, vol. 8 3, pp. 425-432, Sept. 1993.

[57] P.C. Krause, O. Wasynczuk, and S.D. Sudhoff , *Analysis of Electric Machinery.* Piscataway, NJ: IEEE Press , 1995

[58] *TMS320LF2407 Evaluation Module Technical Reference.* Spectrum Digital Inc., Literature no. 504885-001 Rec. C, Stafford, TX, Aug. 2000.

[59] *TMS320C24x DSP Controllers Reference Set, volume-1: CPU, Systems and Instruction Set.* Texas Instruments, Literature no. SPRU161B, Houston, TX, Dec. 1997.

[60] *TMS320C24x DSP Controllers Reference Set, volume-2: Peripheral Library and Specific Devices,* Texas Instruments, Literature no. SPRU161B, Houston, TX, Dec. 1997.

[61] *DMC 1500 Technical Reference,* Spectrum Digital Inc., Literature no. 504915-0001 Rec. C, Stafford, TX, Sep. 2000

VITA

Lei Hao received his Bachelor of Science degree from Shanghai JiaoTong University, Shanghai, China, in July 1991. He then worked as an electrical engineer at the Lanzhou Locomotive Corporation of the Railway Department for four years. In 1995, he started his graduate study at the Institute of Electrical Engineering, Chinese Academy of Science, Beijing, China, and received his Master of Science degree in July 1998. In August 1998, he joined the Ph.D. program of the Department of Electrical Engineering at Texas A&M University, College Station, Texas, with Dr. Hamid A. Toliyat as his advisor, and he completed his Ph.D. degree in December 2002.

Lei Hao is a student member of the IEEE.

He can be reached in c/o

>Dr. Hamid A. Toliyat
>Electric Machines & Power Electronics Laboratory
>Department of Electrical Engineering
>TAMU 3128
>Texas A&M University
>College Station, Texas 77843-3128

Wissenschaftlicher Buchverlag bietet

kostenfreie

Publikation

von

wissenschaftlichen Arbeiten

Diplomarbeiten, Magisterarbeiten, Master und Bachelor Theses
sowie Dissertationen, Habilitationen und wissenschaftliche Monographien

Sie verfügen über eine wissenschaftliche Abschlußarbeit zu aktuellen oder zeitlosen Fragestellungen, die hohen inhaltlichen und formalen Ansprüchen genügt, und haben **Interesse an einer honorarvergüteten Publikation**?

Dann senden Sie bitte erste Informationen über Ihre Arbeit per Email
an info@vdm-verlag.de. Unser Außenlektorat meldet sich umgehend bei Ihnen.

VDM Verlag Dr. Müller Aktiengesellschaft & Co. KG
Dudweiler Landstraße 125a
D - 66123 Saarbrücken

www.vdm-verlag.de

Made in the USA